Praise for
Dude, Where's My Flying Car?

*"The automobile is at all once… the catalyst, the symbol, and the
consequence of the triumph of the individual human spirit."*
—Former State Senator Jack Tate, Colorado

"Dude, Where's My Flying Car?" by author Tim Jackson is a riveting
exposé that takes readers on an unforgettable journey through the
evolution of personal transportation. From the first motorized
carriages of yesteryears to the sleek and sophisticated flying cars of
today, Jackson masterfully navigates through history with a narrative
that is as informative as it is engaging.

With each chapter, Jackson peels back the layers of technological
advancements, the larger-than-life characters and societal changes
that have influenced the development of automobiles. His thorough
research and eye for detail shine through, offering readers a comprehensive look at how each milestone in personal mobility innovation
has been a steppingstone towards the realization of flying cars. The
book is not just a historical account it's a vibrant story of human
ingenuity and relentless ambition.

"Dude, Where's My Flying Car?" is not just for car enthusiasts or
tech geeks, it's a must-read for anyone curious about the journey of
human progress and the future possibilities of transportation.

Tim Jackson also dives into the challenges and setbacks faced
along the way, offering a balanced view that adds depth to
the narrative. He discusses environmental concerns, regulatory
hurdles, and the ethical considerations of a world where cars fly,
grounding his futuristic visions in today's reality.

The book concludes with a forward-looking perspective, pondering the societal impacts of flying cars and how they could reshape our cities, economies, and daily lives. Jackson's optimism is contagious, leaving readers excited about the possibilities that lie ahead.

In "Dude, Where's My Flying Car?", Tim Jackson has crafted an exhilarating read that educates, inspires and entertains. It's a testament to human creativity and a tantalizing glimpse into a future that might just be around the corner. This book is a journey well worth taking, offering a perfect blend of history, technology and dreamy speculation that will leave you looking at the sky in anticipation.

—Mike Marshall, author, speaker, trainer

Dude, Where's My Flying Car?

A fascinating journey into the past, present, and incredible future of America's personal mobility

TIM JACKSON

DUDE, WHERE'S MY FLYING CAR?
by Tim Jackson

Direct requests for permission to
Tim Jackson c/o timwjacksoncae@gmail.com

Books can be ordered for Bulk Sales through the publisher:
timwjacksoncae@gmail.com

Book Design: Nick Zelinger, NZGraphics.com
Editors: Bud Wells, Mike Marshall, Caroline Schomp, Bruce Goldberg

ISBN: 979-8-9903624-0-6 (softcover)
ISBN: 979-8-9903624-1-3 (hardcover)
ISBN: 979-8-9903624-2-0 (e-book)
ISBN: 979-8-9903624-3-7 (audio)
Library of Congress Cataloging in Publication Data on File

Transportation | Bicycles | Motorcycles | Automotive
Aviation | eVTOLs | History | Future

First Edition, Second Printing
Printed in the USA

CONTENTS

PART TWO: THE PRESENT

FOREWORD

At our core, humans covet the ability to get from one place to another, sometimes in a thrilling fashion. We seek the best ways to travel, commute, mobilize and move around our planet. We honor ingenuity, technology, an exhilarating experience or just the ability to save time. For most of us, it is not just the destination we seek, we value the journey as well.

As we stand on the cusp of a new era in personal mobility, it is both elating and daunting to imagine the possibilities that lie ahead. In "Dude, Where's My Flying Car?" the author takes us on a captivating journey through the history, present and incredible future of America's transportation and mobility landscape. It is a canvass yet to fully receive its full color palette—but one we know will be filled with boundless opportunity and daring beauty.

From the earliest days of bicycles, motorcycles and the first private planes to the invention of the automobile—and now the potential for fully autonomous and later flying cars—this book delves into the evolution of how we get from one place to another. It explores the societal, technological and environmental factors that have shaped our transportation systems, and both the challenges and opportunities that lie ahead.

We recognize that there are some among us who want to limit our choices when it comes to transportation and mobility. This book highlights both, athough it seeks to enhance and encourage greater options and enable enjoyment in our travels.

With a keen eye for detail and a passion for innovation, the author paints a vivid picture of what the future of personal mobility could look like. From fully autonomous vehicles to flying cars, the possibilities are endless.

Yet with these advancements come important questions about safety, accessibility and sustainability that must be addressed. In "Dude, Where's My Flying Car?" the author challenges us to think critically, yet optimistically, about the future of transportation and to consider the impacts it will have on our lives, our communities and our planet.

This book is a call to action for all of us to engage in conversations about the future of personal mobility and to work towards creating transportation systems that are safe, efficient and accessible for all.

I hope it inspires you to imagine a future where flying cars are not just a dream, but a reality. And, more importantly, I hope it motivates you to be an active participant in shaping that future.

The journey ahead is sure to be an exciting one. Tim Jackson provides us with a thought-provoking and insightful guide to the road ahead. The bends will be sharp, but the ride will be exhilarating.

– Jason Stein
Host of SiriusXM radio program and YouTube channel,
"Cars and Culture with Jason Stein"
Founder/CEO, Flat Six Media

PART ONE
THE PAST

1

Cars are Essential

Soon they will drive themselves and even fly!

Cars have been a symbol of freedom and independence since their introduction more than 125 years ago. We now see a clear effort by governments globally, as well as transportation planners and advocacy organizations, to make it more difficult for people to buy, own, drive and park cars.

Whether you have noticed it in your own sphere or not, the war on cars is growing and spreading across the United States and around the world and it will undoubtedly make life more problematic for those wanting or needing to commute and enjoy the recreational opportunities that cars make possible.

Cars have revolutionized transportation, allowing people to travel greater distances in shorter periods of time and have played a crucial role in shaping the modern world. Cars have also been instrumental in driving upward economic mobility, allowing people to access better job opportunities and improve their quality of life.

This book is an examination of how cars have facilitated that upward mobility, even though we still can't buy ones that fly.

Other modes of getting around are important as well. Mobility is and should always be a matter of personal choice, although it may not necessarily be the form of mobility that government leaders would choose for us.

Cars first

Our free-market economy has been built on the idea that people should have the opportunity to move up the economic ladder and improve their financial position. Cars have significantly contributed to economic mobility by providing access to improved job opportunities. Prior to the automotive revolution, people generally were limited to working near their homes. The widespread adoption of automobiles allowed people to go farther, faster and more conveniently to take advantage of employment opportunities in other neighborhoods and beyond.

Cars also enabled people to improve their education and training, crucial ingredients for upward economic mobility. For example, students in rural areas often do not have access to universities or training programs within their reasonable local reach. However, with a car, rural residents can travel to the nearest city to attend a university or education and training programs. This not only improves their chances of getting a better job but also increases their earning potential for their lifetime.

Cars have also fostered entrepreneurialism. Cars enable new business owners to conveniently transport goods and services. Cars enable the customer base to reach those new businesses thereby increasing their revenue. This creates an economic domino effect:

more revenue prompting business reinvestment, more employees and faster business growth and development.

Cars also ushered in the freedom to travel and explore the country, sometimes to areas that were previously inaccessible, leading to a greater appreciation of the natural world and the cultural diversity of the United States.

In fact, road trips have become a staple of American culture, providing the means and inspiration for people to explore the vast and diverse landscapes of the country. Travel and exploration have inspired generations of artists and writers who've also relied on cars. The iconic Route 66, for example, is the subject of countless songs, books and movies, capturing the imagination of people around the world. And who can forget Jack Kerouac's *On the Road*? Or the Chevrolet Corvette that starred on television's *Route 66*?

Cars have also played a crucial role in the development of suburban communities. Suburban development boomed in the post-World War II years, fueled in part by the availability of cars. With cars, people could live farther away from city centers and still have access to the amenities and services they desired. The creation of suburban neighborhoods was powerfully symbolic of the American dream.

Cars have helped shape the way we interact with our communities. They have enabled us to access community events, participate in local activities and connect with our neighbors. Cars have also opened volunteer opportunities that contribute to our communities in meaningful ways. Without cars it would be much more difficult to transport food to local food banks, or to drive elderly or disabled members of the community to medical appointments

Better job opportunities, education, training and entrepreneurialism all would be out of reach without cars. The natural world and cultural diversity of the United States are made more accessible by car. Without cars, suburban communities that have defined American life would not have been formed.

Cars are more than just simple transportation or the enabler of personal mobility; they symbolize freedom and independence and have made possible the American dream. As cars advance with far better safety technology, even enabling them to drive or fly themselves and carry us with them, personal mobility and all the benefits it brings, including freedom and independence, are enhanced.

Vehicle miles traveled (VMTs) in the United States has more than tripled in the past 50 years, from 1.1 trillion miles in 1970 to over 3.3 trillion miles by 2020. Even as trillions of dollars have been invested in public transit systems and bicycles have grown exponentially in popularity, ultimately Americans choose to commute by motor vehicle by four-to-one to as much as ten-to-one, depending on which part of the country is tallied. As Americans it should remain our choice as to how we prefer to commute.

Let's see in this Part I how we got where we are in the realm of choice of personal mobility.

2
___

Bicycles

Early bicycle development leads to motorcycles and cars

"The bicycle is a simple solution to some of the world's most complicated problems" —Bill Strickland

One of the most important days of my life was when I learned to ride a bicycle." —Unknown

Thanks to longtime friend and classmate, Joe Clevenger, for teaching me, at age 5, how to ride a bike.

The invention of the bicycle marked a significant milestone in the history of transportation and personal mobility, serving as a precursor to the development of cars. Not only did bicycles introduce the concept of personal mobility, they also laid the groundwork for the advancements that would later shape the automotive industry. To start with, we will explore the early beginnings of bicycles, their influence on the emergence of cars and the involvement of some of the earliest car makers in this transition.

The history of bicycles can be traced back to the early 19th century when the first pedal-powered two-wheeled vehicles were created. These early bicycles, known as "velocipedes" or "dandy horses," were propelled by the rider pushing their feet against the ground. Although they lacked pedals, these contraptions provided a glimpse into the potential of personal mobility and the liberation it could offer.

As the design of bicycles progressed, the addition of pedals in the 1860s revolutionized the cycling industry. This innovation allowed riders to propel themselves forward by rotating the pedals, significantly improving efficiency and speed. The "boneshaker," an early bicycle model equipped with pedals, gained popularity and sparked a new era of transportation.

The rise of bicycles had a profound impact on society, inspiring a wave of enthusiasm for personal mobility. Cycling clubs emerged, creating a community of enthusiasts who sought adventure and freedom on two wheels. Bicycles became a symbol of progress and modernity, transforming the way people perceived transportation and personal mobility.

It is within this context that the early car makers, as well as motorcycle visionaries, found their inspiration. As bicycles gained prominence, engineers and inventors began experimenting with motorized vehicles, cars (or "horseless carriages") and motorcycles. "The principles of balance, steering and propulsion that had been refined in bicycles became the foundation for these new inventions. The knowledge gained from building bicycles provided invaluable insights into the mechanics of transportation, setting the stage for the emergence of cars.

Several of the earliest car makers owe their start to their involvement in the bicycle industry. One such example is Karl Benz, co-founder of Mercedes-Benz. Before venturing into the world of automobiles, Benz was a successful bicycle manufacturer. His experience with bicycles allowed him to understand the importance of balance, weight distribution and efficient propulsion, which he later applied to his groundbreaking inventions in the automotive industry.

Another prominent figure in the early car manufacturing world with ties to bicycles was Henry Ford. Prior to his foray into cars, Ford worked as an engineer early on for the Detroit Bicycle Company, a bicycle manufacturer. His experience in the bicycle business gave him a deeper understanding of the mechanical aspects of transportation, inspiring him to pursue his passion for mass producing cars. Ford's subsequent contributions revolutionized the car industry, making automobiles more accessible and affordable for the masses.

Bicycles played a pivotal role as a precursor to cars, shaping the development of personal mobility and transportation. The early designs and innovations in bicycles laid the foundation for the emergence of automobiles, providing valuable insights into balance, propulsion, and other mechanical principles. Moreover, the involvement of some of the earliest car makers, such as Karl Benz and Henry Ford, in the bicycle industry further highlights the interconnection between these two modes of transportation. The bicycle's influence on the evolution of cars remains a testament to the enduring impact of this humble two-wheeled invention.

Early mass-produced manufacturers and visionary leaders - late 1800s and early 1900s

The late 1800s and early 1900s marked a transformative period in the history of the bicycle industry. During this time, several visionary leaders and mass-produced bicycle manufacturers emerged, revolutionizing personal mobility and transportation as well as leisure activities for people across the globe. One such pioneering company was Schwinn, which played a significant role in shaping the industry. The bicycle contributed greatly to the later-developing motorcycle and new car industries. Here we explore some of the earliest mass-produced bicycle manufacturers and shed light on the visionary leaders behind these companies.

Schwinn

Schwinn was founded in Chicago, Illinois, by Ignaz Schwinn in 1895 and quickly became one of the most prominent bicycle manufacturers of the late 19th and early 20th centuries. The company's commitment to quality and innovation contributed to its success. Schwinn bicycles were known for their durability and reliability, at the time making them highly sought after by consumers. The company's mass-production techniques allowed for the production of affordable bicycles, better ensuring they would be accessible to wide-open markets. Schwinn's ability to adapt to changing consumer demands and introduce new models solidified its continued success throughout the 20th century.

Ignaz Schwinn was a visionary leader who played a pivotal role in revolutionizing the bicycle industry. Schwinn's commitment to innovation and quality allowed his company to flourish. He introduced mass-production techniques to the cycling industry, making bikies more affordable and accessible to the masses. Schwinn's vision for the future of transportation and leisure activities propelled his company to become one of the most influential bicycle manufacturers of its time.

Columbia Bicycles

Another early mass-produced bicycle manufacturer was Columbia Bicycles, founded in 1877 by Albert Pope. Columbia Bicycles gained popularity for its innovative designs. This invention revolutionized the bicycle industry, replacing the previously popular high-wheeler bicycles. Columbia Bicycles' commitment to quality and technological advancements helped it thrive during the late 1800s and early 1900s.

Albert Pope was another visionary leader in the bicycle industry. Pope's innovative designs, such as the chain-driven bicycle, revolutionized the way bicycles were manufactured and used. His commitment to technological advancements and quality craftsmanship helped Columbia Bicycles gain recognition and popularity during this period. Pope's vision for a more efficient and versatile bicycle played a crucial role in shaping the industry.

Humber Cycles

Humber Cycles, founded in 1868 by Thomas Humber, played a crucial role in the mass production of bicycles during this era. Humber Cycles was known for its high-quality craftsmanship and innovative designs. The company's bicycles were highly regarded for their comfort, durability and smooth ride. Humber Cycles became a reputable choice among bicycle consumers.

Thomas Humber was a visionary leader known for his attention to detail and commitment to excellence. Humber's bicycles were renowned for their comfort and smooth ride, setting new standards in the industry. Humber's determination to produce high-quality bicycles earned his company a reputation for excellence. His visionary leadership and dedication to craftsmanship helped Humber Cycles thrive during this transformative era.

Impact and Legacy

The early mass-produced bicycle manufacturers and visionary leaders had a profound impact on the bicycle industry and society as a whole. Their contributions revolutionized personal mobility and transportation, transforming the way people lived and interacted with their environment.

The mass-production techniques introduced by companies such as Schwinn, Columbia and Humber made bicycles more affordable and accessible to a wider range of consumers. This accessibility contributed to the rise of cycling as a popular means of personal mobility, leading to increased mobility and independence for individuals. The bicycle became a symbol of freedom and empowerment, particularly for women who could now travel more freely, and participate in sports and recreational activities.

The visionary leaders behind these companies, such as Ignaz Schwinn, Albert Pope and Thomas Humber, paved the way for future advancements in the bicycle industry. Their commitment to innovation and quality was profound and meaningful.

The bicycle industry has witnessed remarkable growth over the years, with numerous manufacturers contributing to its development and popularity. Important to this history are the current top five bicycle manufacturers, shedding light on their origins, founders and annual production. By examining their journey, we can gain insights into the evolution of the bicycle industry and the impact these manufacturers have had on shaping these other modes of transportation and personal mobility.

Giant

Giant Manufacturing Co., Ltd, commonly known as Giant, is a Taiwanese bicycle manufacturer founded in 1972 by King Liu. Starting as an OEM (Original Equipment Manufacturer), Giant initially produced bicycles for other brands. However, they soon ventured into manufacturing their own bikes under the Giant brand. Today, Giant is recognized as one of the largest bicycle manufacturers globally, producing an estimated 6.6 million bicycles annually. Their commitment to innovation, advanced research and development and carrying a wide range of models has contributed to their success.

Trek

Trek Bicycle Corporation, headquartered in Waterloo, Wisconsin, was founded in 1976 by Richard Burke and Bevil Hogg. Initially, the company focused on producing high-quality steel touring frames. However, they quickly expanded their offerings to include road bikes, mountain bikes, and hybrids. Trek's commitment to quality and technological advancements has made them a leader in the bicycle industry. With an annual production of approximately 1.5 million bicycles, Trek continues to be a major player in the market.

Specialized

Specialized Bicycle Components, commonly known as Specialized, was founded in 1974 by Mike Sinyard. Based in Morgan Hill, California, Specialized has been at the forefront of innovation in the bicycle industry. Specialized was among the first to introduce mass-produced mountain bikes and have since expanded its product range to include road bikes, e-bikes and specialized models for other focuses areas and disciplines. Today, Specialized produces around 400,000 bicycles per year, catering to cyclists of all levels and preferences.

Merida

Merida Industry Co., Ltd, a Taiwanese bicycle manufacturer, was established in 1972 by Ike Tseng. Initially, Merida focused on producing bikes for other brands but eventually transitioned to manufacturing their own bicycles. The company gained international recognition after partnering with professional cycling teams. Merida's commitment to quality, technological advancements, and extensive research and development has made them a prominent player in the industry. With an annual production of approximately 2.2 million bicycles, Merida continues to grow and diversify its product range.

Cannondale Bicycle Corporation

Cannondale Bicycle Corporation, founded in 1971 by Joe Montgomery, is based in Wilton, Connecticut. Originally producing camping gear, the company shifted its focus to bicycle manufacturing in 1983. Cannondale quickly gained attention for its innovative aluminum frames, which offered superior performance and durability. Over the years, they expanded their product line to include road bikes, mountain bikes, and e-bikes. Cannondale's annual production stands at around 250,000 bicycles, emphasizing its commitment to quality over quantity.

3

Motorcycles

How little bicycles grew up to be big motorcycles

William Harley and Arthur Davidson

"Four wheels move the body; two wheels move the soul. Life is short, buy the motorcycle, have a ride, live your dreams. You don't stop riding when you get old; you get old when you stop riding."
—Unknown

Motorcycles conjure up images: Marlon Brando in *The Wild One*; Dennis Hopper and Peter Fonda in *Easy Rider*; Tom Cruise in *Mission Impossible: Rogue Nation*; and, of course, Steve McQueen in *The Great Escape*. These independent, macho heroes (or anti-heroes) exemplify a small slice of motorcycle history that began in the 1800s.

Motorcycle development

While the history of the automobile can be traced back to the 17th century, it's a little harder to pin down exactly when and where the motorcycle began. But bicycles and synchronicity were definitely the foundations as inventors in America, Europe and Asia borrowed from and built on one another's contributions.

The Americans

American inventor Sylvester Roper attached a two-cylinder, coal-fired, steam-powered engine to a velocipede—that vehicle with the huge front wheel—in what is assumed to be 1867. Arizonan Lucius Copeland built a smaller, steam-powered velocipede in 1881 that achieved 12 mph.

Charles H. Metz, founder of the Waltham Manufacturing Co. in Massachusetts, first used the term "motor cycle" in an advertisement for his Orient motor-powered cycle in 1899. The Orient won the first recorded American motorcycle speed contest on July 31, 1900.

Landmark American contributions to motorcycling began in the 1900s. Indian Motorcycle adopted a four-stroke engine, developed by the French DeDion-Buton, in 1901. Indian had a hit with its single model and was producing more than 20,000 units by 1913.

Two years later, William Harley and partners Arthur and Walter Davidson collaborated to found Harley-Davidson out of a wooden shack in Milwaukee, Wisconsin. By 1905 they were producing a commercially viable bike, developing the first dual-cylinder engine in 1909. Variations of that iconic engine are still part of the Harley lineup, sold in 67 countries—and Harley is the largest motorcycle manufacturer in the world.

Indian and Harley-Davidson remained the only two American motorcycle manufacturers until 1953 when Indian closed and British maker Royal Enfield took over the Indian name. Indian made a comeback in 2013 under the aegis of Polaris, a maker of ATVs and snowmobiles.

The French

Pierre Michaux attached pedals to a primitive velocipede in 1861, taking the two-wheeler to popularity. According to Louis Schafer in *American Motorcyclist,* "… aching legs prompted imaginative minds to search further for a better power source." Michaux's son, Ernest, added a small steam engine to his father's "boneshaker" bicycle.

Inventor Guillaume Perreaux patented a commercial alcohol-fired, single-cylinder steam engine affixed to Michaux's bicycle in 1868 with "… a maximum speed of about 9 mph. However, it also used up its water supply quickly and was dangerous to the rider in the slightest mishap or fall." It didn't have brakes.

The DeDion-Bouton engine—a lightweight, four-stroke engine that produced 1.3 hp—was introduced in Paris in 1895 and became popular with moto-bicycle builders, including Indian. Peugeot Motocycles exhibited a motorcycle powered by an engine invented by DeDion-Buton in 1898. It continued to build motorcycles until WWII. After the war, while still building some motorcycles transitioned more toward mopeds and scooters.

The Italians

As with clothes, shoes and fast cars, Italians have created some standout motorcycles. "What is a motorcycle from Italy? In the first place, it is speed. Second, it is a high level of exterior elegance," as one source put it. The best-known brands include Ducati, MVAgusta and Aprilia.

Ducati produced radio transmitters until World War II. Post-war, it fulfilled a need for cheap engines attached to bicycles with the Ducati Cucciolo, followed by the two-hp Ducati 60 in 1949 and bigger bikes in the 1950s. Street racing fueled the desire for more bikes. Ducati has continued to produce motorcycles for both streets and tracks. Ducati is famous for introducing the "naked bike," or "standard," with no wind-deflection and exposed chassis.

Its website enthuses: "...the passion remains, for a true, unique, adrenaline-filled experience, like a red motorcycle hurtling down the roads and racetracks throughout the world."

Aircraft maker MVAgusta rebuilt post-WWII to manufacture motorcycles. Racing enthusiast Count Domenico Agusta applied racers' characteristics to production bikes. The company has changed hands several times, and is now owned by multinational investors. It continues to manufacture high-end road and racing bikes.

Aprilia, formed after WWII to manufacture bicycles, moved into scooters and motorcycles. It's built renowned racing and motocross bikes. It continues to excel with its racers after acquisition by Piaggio, which is better known for its Piaggio and Vespa scooters.

The Germans

Gottlieb Daimler (Mercedes-Benz) and Wilhelm Maybach attached their first gas-powered combustion engine to a traditional bicycle, called the Daimler Reitwagen, in 1885. The Reitwagen's ½-hp engine got up to 7 mph. It is credited as the first modern motorcycle.

Hildebrand & Wolfmuller, the first to factory-build motorized two-wheelers, coined the name "motor-cycle" (in German: motorrad) beginning in 1894.

NSU—another bicycle maker—debuted its first motorcycle in 1901. It achieved racing fame, breaking many world records, including a 211-mph World Land Speed Record at Bonneville in 1956. During the '30s NSU was the largest motorcycle manufacturer; by the mid-'50s it produced 350,000 units. NSU turned more to automobile engine development and ceased motorcycle production in 1963.

BMW was banned from aircraft engine manufacturing post-World War I. Its first motorcycle—the R32—was introduced in 1923. A consistent innovator, BMW's original flat-twin boxer engine configuration is still used along with other designs and it introduced the first anti-lock brakes. BMW sold more than 175,000 units in 2021.

The British

British motorcycle history is all about mixing, matching and consolidation. In the 1930s there were more than 80 makers of British motorcycles, and over the years there have been 280 identified British brands, though *Shooting Star: The Rise and Fall of the British Motorcycle Industry,* reported British production peaked in 1928 at 147,000 machines.

The Birmingham Small Arms Company Ltd. (BSA), incorporated in 1861, sold guns and bicycles. BSA transitioned to motorcycles with the BSA 3 1/2 hp. BSA eventually became the largest motorcycle producer in the world by buying out many of its well-known competitors.

Ariel produced its first motorcycle in 1902. Acquired by Components Ltd. that year, it continued producing a three-speed, two-stroke bike, the Arielette, which it made until the beginning of World War I. After the war it continued building motorcycles, becoming part of BSA until its demise in 1965.

Triumph, founded in 1884 to sell bicycles, produced a motorcycle in 1902—a one-cylinder machine with a Belgian engine; 500 were sold that year. Triumph was selling 3,000 units annually by 1909. By 1930, it sold 30,000 units in eight models. Triumph was sold to Ariel in 1935 and to BSA in 1951.

Norton Manufacturing, founded in 1898 to make bicycle parts, soon was assembling motorcycles with other makers' parts. Norton became known for racing motorcycles. The company failed in 1913 but was resurrected. It absorbed BSA-Triumph in the mid-'70s, failed again and now is in the hands of India's TVS Motor Co.

Royal Enfield Motorcycles began building bicycles in 1882, making parts for others before transitioning to motorcycles in 1901 as The New Enfield Cycle Company Ltd. It was sold twice before being absorbed by BSA in 1907. The Royal Enfield Bullet, still manufactured by Enfield of India, is considered the longest-lived motorcycle in history.

Wartime

The U.S. Army reportedly used Harleys to hunt Pancho Villa in the Mexican revolution. Motorcycles played a critical role in both world wars. Motorcycles were better than horses for delivering dispatches to the front lines in World War I and a machine gun could be mounted on them. Harley-Davidson sent 20,000 units to Europe for Army use in World War I. Indian contributed most of its production during the Great War.

British forces relied on some 30,000 Triumph Model Hs while Norton contributed more than 100,000 WD161Hs. Peugeot offered more than 1,000 bikes to support French troops.

World War II also was fought with motorcycles. The modified Harley-Davidson WL maneuvered well and was sturdy—70,000 were produced for the war effort, including thousands that ended up in Russian allie's hands. Britain ordered more than 126,000 BSA M20s; thousands of lightweight Royal Enfield Flying Fleas were parachuted behind enemy lines.

As good as these motorcycles were, the BMW R71 may have been better. Used extensively in Europe, it excelled in North Africa for its resistance to desert grit.

World War II heralded post-war motorcycle popularity. Returning veterans saw how these wartime workhorses could translate into excitement and fun on America's roads, especially as bigger and faster bikes were produced and the highway system was developed.

The Japanese

Japan's motorcycle industry blossomed after WWII when cheap transportation was necessary for rebuilding. The Japanese since have become motorcycling innovation and competition juggernauts.

Suzuki, Kawasaki and Yamaha motorcycles grew out of existing companies. Honda Motor Co., incorporated in 1948, produced a 98cc bike in 1949, the Dream, or Model D, followed in two years with the four-stroke Dream E. Honda kept building bigger, faster bikes, and became a racing powerhouse. The renowned Gold Wing was introduced in 1974. Honda is now the largest motorcycle manufacturer in the world.

Suzuki got into motorcycles in 1951 by clipping a motor to a bicycle. Over the years it has gained fame for its off-road, sport and racing bikes.

Kawasaki Motorcycles were part of a huge industrial conglomerate that acquired Meguro motorcycles in 1963 and the next year produced its first Kawasaki. It established a U.S. presence in 1966 with small two-cylinder Omegas. Kawasaki quickly gained a reputation for endurance racing and off-roading.

In 1955, Yamaha's first bike, the Red Dragonfly, won two races. Yamaha imported two streetbikes into the U.S. and also began competing on the world racing circuit at the 1958 Catalina Grand Prix, the first time a Japanese motorcycle raced in America. The 1975 YZ Monocross set the tone for future motocross competitions.

The future

U.S. motorcycle use is growing: 12,231,000, a 2 million + increase over 2014 according to the latest data (2018). The Motorcycle Industry Council reports shifting demographics:

- Households owning motorcycles: from 6.94% in 2014 to 8.02% in 2018.

- Gender: Male owners – 81%, with growing female owners; 22% GenX and 26% female Millennial owners.
- Age: Getting older. Median age of motorcycle owners in 2012 was 45; in 2018 it was 50. Retirees equal almost a quarter (24%).
- Education: College graduates increased between 2012 (17%) and 2018 (24%).

U.S. motorcycling is dwarfed by other countries. According to bicyclehistory.net, "Daily over 200 million motorcycles are in use all around the world ... Leading countries with greatest motorcycle use are India (37 million motorcycles/mopeds) and China (34 million motorcycles/mopeds)."

In dense, urban areas of developing countries, motorcycles and their smaller brethren—scooters—are less expensive and more maneuverable on narrow and crowded streets than cars. In the United States, large cruisers are gaining popularity, along with three-wheel motorcycles (trikes for adults!). Some of these larger vehicles are more expensive than automobiles and as the data indicates, older people with more disposable income and a yen for the independence associated with motorcycling, will be hitting the road on two (or three) wheels.

Motorcycle glossary

There is a motorcycle for almost every use:

- *Street bikes* – for riding on paved roads. Usually, street bikes have 125 cc or larger engines, and light tread on their tires.
- *Standards/naked bikes/roadsters* – general purpose bikes designed for upright riding without windscreens.
- *Cruisers* – defined by Harley-Davidson with forward foot placement, higher handlebars and lower ground clearance.

Especially good for low to moderate speeds. Power cruisers have more power and higher ground clearance.

- *Sport bikes* – built for more speed, acceleration and handling on paved roads. They position riders' legs closer to the body and riders lean over more to reach the controls.
- *Touring* – built for driving longer distances in an upright, relaxed posture. Generally have large-displacement engines, bigger fuel tanks, windscreens, and more luggage capacity.
- *Dual-sport* – built to be street legal but can also go off-road. A subset is Supermoto, designed to bridge road and track racing and motocross, but is still street legal.
- *Utility* – adapted to perform specific jobs. These include deliveries, police work and other specialized uses.
- *Off-road* – for riding on unpaved surfaces with lighter weight, better ground clearance, flexibility and knobby tires.
- *Motocross* – for racing on closed off-road tracks with obstacles like hills and jumps.
- *Enduro* – built for events that can last days and may include both road and off-road. A subset is the Rally, with a larger fuel tank to accommodate very long distances like the Paris-Dakar rally.
- *Trials* – built for a specialized kind of competition. Riders stand, rather than sit, so these bikes are built for good balancing and precision shifting.
- *Track racers* – self-explanatory. They are built for speed with no brakes, no rear suspension and only two gears.
- *Trail/Dirt* – for general off-roading.

From a personal mobility standpoint, both bicycles and motorcycles have developed to be key components of the personal mobility mix. And, it is my belief, they always will be.

4

Aviation

The launch and evolution of aviation as personal mobility

Orville and Wilbur Wright at Kitty Hawk, 1903

"The desire to fly is an idea handed down to us by our ancestors who looked enviously on the birds soaring freely through space."
—*Wilbur Wright*

"The airplane stays up because it doesn't have time to fall."
—*Orville Wright*

The Wright Brothers, Orville and Wilbur, were pioneers in the field of aviation, revolutionizing the way we perceive and travel through the skies. Born in the late 19th century, the brothers hailed from a small town called Dayton, Ohio. However, it was not their place of birth that defined their destiny, but rather their insatiable curiosity and passion for innovation. Before venturing into the realm of aviation, the Wright Brothers developed a keen interest in bicycles.

In the late 1800s, bicycles were becoming increasingly popular, and the brothers saw an opportunity to combine their mechanical skills with their love for cycling. They opened a bicycle sales and repair shop in Dayton, where they honed their engineering skills and gained a deep understanding of mechanics and aerodynamics. Their fascination with mechanics didn't stop at bicycles. Orville and Wilbur were always drawn to the latest technological advancements, and as the turn of the century approached, their attention shifted towards the emerging field of automobiles.

The Wright Brothers opened a car dealership and repair shop alongside their bicycle business, further expanding their knowledge of engines and machinery. However, it was the allure of flight that truly captivated the Wright Brothers. Inspired by the works of aviation pioneers such as Otto Lilienthal and Octave Chanute, the brothers became determined to conquer the skies. They delved into the study of aeronautics, devouring books and articles on the subject and conducting their own experiments.

In their quest to build a successful flying machine, the Wright Brothers were drawn to Kitty Hawk, a remote coastal area in North Carolina. The location offered several advantages that made it an ideal testing ground for their aircraft. The consistent strong winds, sandy dunes, and the lack of obstructions provided an opportunity for the brothers to perfect their designs and achieve controlled flight.

The Wright Brothers' choice to build their early planes and fly them at Kitty Hawk was not arbitrary. They recognized that the area's conditions closely resembled those of the ideal flight environment. By conducting their experiments there, they were able to gather crucial data and make incremental improvements to their aircraft.

On December 17, 1903, the Wright Brothers achieved their ultimate goal. In the dunes of Kitty Hawk, their aircraft, the Wright

Flyer, took off and remained airborne for 12 seconds, covering a distance of 120 feet. This historic event marked the birth of powered flight and forever changed the course of human history.

Orville and Wilbur Wright's interest in bicycles and cars, along with their insatiable curiosity, led them to pursue the dream of flight. Their meticulous study of aeronautics and their choice to test their aircraft at Kitty Hawk were instrumental in their success. Through their determination and ingenuity, the Wright Brothers paved the way for aviation as we know it today, leaving a lasting legacy that continues to inspire generations of aspiring aviators.

From the Wright Brothers to modern personal aircraft, the invention of the airplane in 1903 marked a significant turning point in human history.

This groundbreaking achievement paved the way for the development of the aviation industry as we know it today. The aviation industry has revolutionized global mobility, transportation and connectivity. Here we explore the evolution of the aviation industry since the Wright brothers' first flight, highlighting prominent aircraft manufacturers and their contributions to the field with a particular focus on small, personal airplanes.

Essentially a larger and sturdier version of the 1902 glider, the only fundamentally new component of the 1903 aircraft was the propulsion system. With the assistance of their bicycle shop mechanic, Charlie Taylor, the Wrights built a small, twelve-horsepower gasoline engine.

Taylor was a brilliant, self-taught mechanic who made a crucial contribution to the Wright brothers' invention of the airplane. When a suitable engine was not available for the first flight, he designed and built the first successful airplane engine in only six weeks.

Taylor began working in the Wrights' bicycle business in 1896, and he played an important role in their flying experiments for several years. He also served as chief mechanic for the first transcontinental flight, from New York to California, achieved in

1911 by Cal Rodgers. Charlie Taylor's birthday, May 24, is celebrated as Aviation Maintenance Technician Day.

Development of Aviation

Following the Wright brothers' first successful flight, the aviation industry experienced rapid growth and advancement. The early 20th century witnessed the emergence of several renowned aircraft manufacturing companies that played a pivotal role in shaping the industry.

Companies such as Boeing, Lockheed Martin, Airbus, Piper, Beechcraft, Mooney, Cirrus and Cessna have become synonymous with aviation excellence.

Boeing: Founded in 1916, is one of the most influential and successful aircraft manufacturers in the world. Initially, Boeing focused on producing military aircraft during World War I. However, the company expanded its portfolio to include commercial planes in the post-war period. Boeing's iconic aircraft, such as the 747 Jumbo Jet and the 787 Dreamliner, have redefined air travel by introducing unprecedented levels of comfort, efficiency and range.

Lockheed Martin: Established in 1995 through the merger of Lockheed Corporation and Martin Marietta, is renowned for its production of military aircraft. The company has been at the forefront of technological advancements, manufacturing innovative fighter jets and reconnaissance planes. Lockheed Martin's F-35 Lightning II, a fifth-generation multirole fighter, represents a pinnacle of engineering and has become a symbol of modern military aviation.

Airbus: Founded in 1970 as a consortium of European aerospace companies, it has emerged as a fierce competitor to Boeing in the commercial aviation sector. With its headquarters in Toulouse, France, Airbus has gained global recognition for its wide range of aircraft models, including the A320, A380 and A350. These aircraft are renowned for their fuel efficiency, passenger comfort and advanced avionics systems.

Cessna, 1927: While Boeing, Lockheed Martin and Airbus dominate the commercial and military sectors, Cessna has made significant contributions to the development of small, personal airplanes. Cessna has focused on manufacturing light aircraft, including single-engine propeller planes and business jets. The Cessna 172 Skyhawk, introduced in 1955, remains one of the most popular and widely used aircraft in the world, serving as a primary training aircraft for pilots, including myself.

Piper, 1927: Headquartered in Vero Beach, Florida, Piper has maintained a steady production volume over the years, catering to both general aviation and training markets. The total number of planes produced as of 2021, was over 150,000 aircraft, showcasing its enduring presence in the small airplane market.

Beechcraft, 1932: Headquartered in Wichita, Kansas, Beechcraft has varied production volumes throughout its history, adapting to market demands. The number of planes produced by Beechcraft has surpassed 54,000, including popular models like the Bonanza and King Air series.

Mooney, 1929: Headquartered in Kerrville, Texas. Mooney has experienced fluctuating production volumes over the years, influenced by market conditions. Mooney has produced around 11,000 aircraft, known for technological and efficient performance.

Cirrus, 1984: Founded and headquartered in Duluth, Minnesota, Cirrus has seen consistent growth in production volumes, driven by the popularity of their technologically advanced aircraft. Cirrus has produced over 8,000 aircraft, renowned for their composite construction and innovative safety features, including usable parachutes if a plane loses power in the air. So far, Cirrus is the only major aircraft manufacturer to perfect the parachute safety system for its mass-produced planes.

These are just a few examples of the prominent small airplane manufacturers that have left a meaningful impact on the market. Cessna, Piper, Beechcraft, Mooney and Cirrus, among others, have contributed significantly to the aviation industry, providing pilots with reliable, efficient and innovative aircraft. Their rich histories, home office locations, production volumes and total number of planes produced demonstrate their enduring presence and importance within the small airplane market.

When growing up in Northwest Missouri, I was blessed with access to small private planes available for pilot in command training. A friend of mine, later father-in-law, bought a 1957 Cessna 172, with plans to get his pilot license and to make it available for all of his kids who wanted to learn to fly. With four sons, two daughters and soon a son-in-law, his investment in airborne mobility was a gift from God for those of us who couldn't have found the bandwidth to do that without him and his plane.

The plane felt old even in 1975 and 1976 when I was learning to fly. It was built the same year I was born. After starting lessons, it was only three months before I was able to fly solo and another six months before I was fully licensed as a Private Pilot, Single-engine land, Visual Flight Regulations (VFR). I was really the only one of six family members that completed the process to become licensed.

Flying is a fun sport and expedient way of gaining mobility, under the right circumstances. It can be costly. In some cases, it can be inconvenient. Though as the nation and world looks to getting people from one place to another, the air seems to be the last, extremely promising, frontier. The Wright Brothers were breaking ground (or in this case air) in more ways than one might know. Orville and Wilbur Wright could be the most meaningful frontrunners in the early adaptation of flying cars (passenger drones, VTOLs). The aviation industry has come a long way since the Wright brothers' groundbreaking first flight at Kitty Hawk.

Companies such as Boeing, Lockheed Martin, Airbus, Piper, Beechcraft, Mooney, Cirrus and Cessna have played instrumental roles in shaping the industry and pushing the boundaries of technological innovation. While Boeing, Lockheed Martin and Airbus continue to dominate the commercial and military sectors, smaller planes and airplane makers contribute to additional options for people seeking speedier choices toward their personal mobility.

5

Mercedes-Benz

Daimler and Benz – Early automotive pioneers

Karl Benz

"The best or nothing, in the past, present and future."
—Karl Benz

Germany led Europe and the world in development of automotive industry with Daimler-Benz

The automobile industry has come a long way, baby. You can say that as it relates to how far we have come since the inception of the auto industry very late in the 19th century. A number of pioneers played a key role in the development of the auto industry. Two of the most important names associated with it early on were Karl Benz and Gottlieb Daimler.

Each of these two important individuals founded their respective companies in the early 19th century and their legacy lives on very proudly to this day. This chapter is intended to cover the founding of Daimler and Mercedes-Benz and their historic and central connection to the start of the worldwide automobile industry in its earliest days.

The origins of Daimler

Gottlieb Daimler was born in Schorndorf, Germany, in March of 1834. Daimler was a truly brilliant engineer. He worked for a number of companies in the early years of his career. In 1882, Daimler founded his own company, then formed Daimler-Mototoran-Gesellsch (DMG) with his business partner Willhelm Maybach. The company was based in Stuttgart, Germany, and was focused on the development of internal combustion engines.

Daimler's first major success came in 1885, when he developed a high speed engine that could be used in a variety of applications. This engine was particularly well-suited for use in automobiles which were getting popular during this period of time. Daimler's engine was smaller and lighter than most others of the day which, seemingly, made it perfectly ideal for use in cars.

The next few years turned into a period of rapid growth for DMG. The Daimler/Maybach refined engine designs were soon being used on everything from boats to aircraft, though it was the automobile that would ultimately become DMG's most important and timely project.

In 1892, DMG introduced its first automobile, the Daimler Motor Carriage. This vehicle was a major milestone in the history of the automobile, as it was truly the first in the modern sense of the word. It was powered by a four-stroke, internal combustion engine and it had a number of features still found in cars today, including a steering wheel and a differential in the drivetrain.

The origins of Mercedes Benz

While Daimler was busy building his company, another pioneer in the young auto industry was making his mark. Karl Benz was born in Muhlberg Baden, Germany, in 1844. Benz was an engineer who had a passion for designing and building machines. In 1883 he founded Benz & Cie based in Mannheim, Germany. Benz, viewed by many as the father of the car, has also been referred to as the father of the automobile industry.

Benz's first major success came in 1885 when he developed a three-wheeled vehicle, powered by a gasoline engine. This vehicle, the Benz Patent Motorwagen, was widely regarded as the first true automobile. Like Daimler's motor carriage, it was powered by a four-stroke internal combustion engine.

Over the next few years, Benz continued to refine his designs and his company grew rapidly. By the turn of the century, Benz & Cie was one of the largest automobile manufacturers in the world.

The merger of Daimler and Benz

Despite their success, DMG and Benz & Cie faced significant challenges in the early years of the 20th century; competition in the automobile industry was fierce. There were concerns about the long-term viability of both companies.

In 1924 the two companies announced they would merge, forming a new company called Daimler-Benz AG. The merger was a major milestone.

Overall, the 1920s was a time of great change in the automotive industry. Mercedes-Benz was at the forefront. In 1926, the company introduced the Mercedes-Benz 8/38 PS, a car that was both stylish and powerful. The 8/38 PS was powered by a four-cylinder engine that produced 38 horsepower with advanced features that included hydraulic brakes and a four-speed gearbox.

The name 'Mercedes' in the car name Mercedes-Benz comes from the daughter of Emil Jellinek, a businessman and racing enthusiast who played a significant role in the early development of the Mercedes brand. The cars were named after his daughter, whose name was Mercedes Jellinek. Besides being actively engaged in leadership of the car manufacturer, he was also an early new car dealer in Austria

Emil Jellinek, (known after 1903 as Emil Jellinek-Mercedes), was an automobile entrepreneur responsible in 1900 for commissioning the first modern automobile, the recedes 35hp. Jellinek created the Mercedes trademark in 1902.

Whenever people talk about Mercedes-Benz early motorsport success, most reference the Silver Arrows that dominated the motorsport world in the late 1930s. However, well before that, there was a line of White Arrows emerged in 1927, winning many hill climb events, two Nurburgring GPs (Grand Prix, 1928 and 1931), one Irish GP (1930) and the most challenging road race in the world, Mille Migila (1931). The White Arrows broke the domination by Italian and French sports car makers and gave German the first taste of motorsport success. That was the Mercedes S / SS / SSK / SSKL series.

The S / SS / SSK / SSKL series was designed by Ferdinand Porsche when he was serving as technical director of Daimler, just before founding his own consultant company. Work was started based on the existing Mercedes 630K, which was a luxury touring car powered by a 6.3-liter supercharged straight-6.

Porsche gave the new S (Sport) model a lower chassis for better handling and bored out the engine to 6.8 liters, taking its supercharged horsepower to 180. As before, the Roots-type supercharger was engaged by flooring down the throttle, otherwise the engine ran off-boost and produced up to 120 hp.

Therefore, the S was officially called 26/120/180S, where the numbers represented its tax horsepower, off-boost horsepower and supercharged horsepower respectively. Another superior feature to the 630K was a four-speed gearbox, which could take the car to 110 mph.

Mercedes-Benz continued to innovate in the 1930s with the introduction of the Mercedes-Benz 540K. It was truly a masterpiece of engineering with a sleek and stylish design that was unmatched by any other car of the time. The 540K was powered by a supercharged eight-cylinder engine that produced an impressive 180 horsepower, making it one of the best performing cars of its era. The 540K was also equipped with a hydraulic suspension system and a four-speed gearbox.

The 1940s were a difficult time for Mercedes-Benz, as the company was forced to focus on producing military vehicles during World War II. However, after the war, Mercedes-Benz quickly bounced back with the introduction of the 300 SL, a true masterpiece of design, with a sleek and aerodynamic body inspired by the company's racing cars. The 300 SL was also equipped with a range of advanced features, including a fuel-injected engine that produced 215 horsepower, making it one of the fastest cars of its time.

The 1950s saw Mercedes-Benz continue to innovate with the introduction of the 190 SL and the 300 SLR. The 190 SL, a stylish and affordable sports car designed to appeal to a wider audience, was powered by a four-cylinder engine that produced 105 horsepower, making it a relatively quick car for its time. The 300 SLR, on the other hand, was a true racing car that was designed to compete in some of the most grueling races in the world. The 300 SLR was powered by a fuel-injected eight-cylinder engine that produced an impressive 310 horsepower, one of the most powerful cars yet.

Mercedes-Benz continued in the 1960s to push the boundaries of what was possible in the automotive industry with the introduction of the 600, a masterpiece of engineering with a range of advanced features unmatched by any other vehicle of its time.

The 600 was powered by a massive six-liter V8 engine of 300 horsepower, and a four-geared automatic transmission.

In the 1970s, Mercedes-Benz continued to innovate with the introduction of the S-Class. This car was a true luxury sedan with a range of advanced features designed to provide the ultimate driving experience. The S-Class was powered by a range of engines, including a fuel-injected V-8 that produced up to 245 horsepower.

The 1980s saw Mercedes-Benz introduce the 190E, a sports sedan powered by a range of engines, including a fuel-injected four-cylinder engine that produced up to 190 horsepower. The car was also equipped with a five-speed manual transmission.

During the 1990s, Mercedes-Benz continued to expand its product line and improve its technology. In 1991, the S-Class flagship model featured airbags and traction control, as well as a powerful V-8 engine and a sleek, aerodynamic design. Mercedes-Benz introduced the C-Class in 1993, which was a compact luxury sedan designed to compete with other high-end vehicles in the market. The C-Class was a huge success, and it helped to establish Mercedes-Benz as a major player in the luxury car market.

A notable achievement for Mercedes-Benz in 1995 was the introduction of the E-Class, a midsize luxury sedan that featured a more modern design and advanced technology, including a multilink suspension system and a new electronic stability control system. However, the 1990s also presented some challenges for Mercedes-Benz. The company faced increased competition from other luxury car brands, such as BMW and Audi, and it struggled to keep up with the changing demands of consumers. Additionally, Mercedes-Benz faced criticism for some of its design choices, such as the use of plastic parts in its cars.

Despite these challenges, Mercedes-Benz remained a top player in the luxury car market, and it continued to innovate and improve its products. In the early 2000s, the company introduced several new models, including the SLK-Class, which was a two-seater roadster that combined sporty performance with luxury features.

In 2006, Mercedes-Benz introduced the CLS-Class, which was a four-door coupe that was designed to appeal to younger, more style-conscious consumers. The CLS-Class featured a sleek, streamlined design and a powerful V-8 engine, and it quickly became a popular choice for buyers wanting a luxury car both practical and stylish.

Another major achievement for Mercedes-Benz during this time was the introduction of the SLS AMG in 2010. The SLS AMG was a high-performance sports car that featured a powerful V-8 engine and a lightweight aluminum body. It was designed to compete with other high-end sports cars, such as the Porsche 911 and the Audi R8, and it quickly became a favorite among car enthusiasts.

The 2010s presented other challenges for Mercedes-Benz, increased competition from Tesla, which was disrupting the traditional luxury car market with its electric vehicles. Additionally, Mercedes-Benz faced criticism for some of its design choices, such as the use of large grilles and overly complicated infotainment systems.

Despite these challenges, Mercedes-Benz continued to innovate and improve its products. In 2014, the company reintroduced the C-Class sedan. Its popular compact luxury car, featured a more modern look, advanced safety features and improved fuel efficiency.

In 2015, Mercedes Benz introduced the GLE-Class, which was a midsize luxury SUV, combining comfort, performance and versatility. The GLE-Class featured a range of advanced features, such as a 360-degree camera system and a self-leveling air suspension, and it quickly became a popular choice for families and adventurers alike.

Mercedes-Benz during this time introduced the AMG GT, a high-performance sports car designed to compete with other high-end sports cars, such as the Porsche 911 and the Audi R8. It featured a high-performance V-8 engine, a lightweight aluminum body and a range of advanced performance features, such as active aerodynamics and a torque-vectoring system.

In recent years, Mercedes-Benz has continued to innovate and improve its products. In 2018, the company introduced the A-Class sedan, which was a new entry-level luxury car that was designed to appeal to younger, tech-savvy consumers. The A-Class featured a range of advanced technological features, such as a digital instrument cluster and switches.

Mercedes-Benz is also now rolling with some exceptional electric vehicles (EVs) in newly introduced EQ line of zero-emission cars. These include EQB, an SUV, EQE, both sedan and SUV, and EQS, both sedan and SUV. The reviews these vehicles are receiving are strong, though the most frequent criticism is that they are priced much higher than comparable in-market EVs. That may be an issue for Mercedes-Benz going forward.

6

Oldsmobile

Ransom Eli Olds – earliest automotive industry pioneer

Ransom Eli Olds

"The man who built the first Oldsmobile was a genius."
—David Buick

R ansom Eli Olds was an American entrepreneur and inventor who left a significant impact on the automotive industry with his innovative ideas and inventions. Olds was born in Geneva, Ohio, the youngest son of a blacksmith and patternmaker. Olds' parents moved the family to Cleveland, Ohio, when Olds was still a boy. The Olds family eventually settled in Lansing, Michigan, where he attended high school before dropping out so he could work full-time at the family company, P.F. Olds & Son.

Ransom Eli Olds' contributions to the automobile industry are numerous, including founding Oldsmobile in 1897, which was later sold to General Motors in 1908. However, Olds' entrepreneurial spirit was not confined to the new passenger car industry as he went on to start the REO Motor Company, an early yet prominent manufacturer of trucks and buses.

On March 9, 1901, the Olds Motor Works factory burned to the ground. Only one model, the little Curved Dash Runabout, was saved from the intense flames. Ransom Olds claimed it was the fire that made him select the Runabout, from among his many other models, to put into production. His biographer later questioned the veracity of the reason the Runabout was selected. He pointed to an Olds advertising blitz that had already led to more than 300 Curved Dash Oldsmobile orders even before the fire took place. The biographer said, "Olds did not need the one rescued car from which to reconstruct the plans and patterns for the Runabout."

Later that year, Olds had his company's test driver Ray Chapin drive a Curved Dash Oldsmobile Runabout to the second annual New York Auto Show. Along the way, Chapin opted to drive up along the Erie Canal tow path to escape the mire of New York state roads. After eight days of driving, Chapin finally reached the Waldorf Astoria Hotel, but was turned away at the door. His mud-spattered attire was so objectionable he was sent to the servants' entrance in the back of the hotel.

During the auto show Olds pushed hard to make sales. When one franchised Oldsmobile dealer offered to purchase 500, Olds retorted, "I would like to see you make this order for a thousand cars. Then the public would drop its jaw and take notice." The deal was signed, and though the dealer ended up selling only 750 to the public, it was the original number that everyone remembered.

The Curved Dash Oldsmobile sold for relatively affordable $650, equal to $21,172 today. About 600 were sold in 1901, about 3,000 in 1902 and at least 4,000 in 1904. It was this car, rather than Henry

Ford's Model T, that was the first mass-produced, low-priced American motor vehicle.

Ransom Eli Olds founded the REO Motor Company in 1904 after he left Olds Motor Company due to disagreements with the board of directors. The new company was named after Olds himself, with the initials, REO, standing for his name. The company initially produced cars, but it was not long before he switched his focus to trucks which proved to be a more profitable business.

REO trucks quickly found early popularity across the United States, given their strong reputation for reliability and durability. The company's first truck, the REO Speed Wagon, was introduced in 1915 and became a staple of the American economy. During World War I, the US Army used REO trucks extensively, which helped establish the brand's reputation for toughness and dependability.

In the 1920s, REO expanded its product line to include buses, which were sold under the REO and Diamond REO brands. These buses were used for public transportation and school buses, as well as long-distance travel. REO buses quickly gained a reputation for being some of the most reliable and comfortable vehicles on the road. In fact, Prevost, a Montreal-based bus manufacturer, used REO engines and chassis for many of its early bus models. Prevost is still one of the top luxury bus makers in the world, well-known for their over-the-top creature comforts.

When Ransom Eli Olds passed away in 1950, his legacy continued through the REO Motor Car Company. The company remained in business until the early 1990s, producing a variety of trucks and buses that helped drive the American economy forward.

Interestingly, some of Ransom Olds' legacy extends beyond the automotive industry and into the world of music. In the 1960s a rock band from Champaign, Illinois, took the name REO Speedwagon, inspired by the classic REO Speed Wagon truck. The band became one of the most successful rock bands of the 1980s with its hits like "Can't Fight This Feeling" and "Keep On Loving You."

Similarly, in the 1970s, a country rock band from Pittsburgh, Pennsylvania, took the name Diamond RIO inspired by the REO truck brand Diamond REO and released several albums in the 1970s and 1980s. Their music is still remembered by fans of classic rock and country music, especially for their hit song "Meet in the Middle."

Although Ransom Eli Olds did not have direct involvement in the formation of these bands, his legacy as an innovator and entrepreneur who inspired generations of musicians and fans such as the REO Speedwagon and Diamond RIO bands are just two examples of how Ransom Eli Olds legacy continues to influence American culture to this day.

7

Cadillac

The car company named for a French explorer who founded Detroit

Henry LeLand

"The Cadillac is the car of cars. No other automobile can match its combination of beauty, luxury and performance."
—Henry Leland, Founder of Cadillac

The Cadillac Motor Company is a historically significant American automaker that has been around for more than a century, producing some of the most luxurious and innovative vehicles in the industry. The company was founded in 1902 by Henry M. Leland, a mechanical engineer and entrepreneur, who had previously worked for the automotive industry pioneer, Ransom E. Olds, founder of Oldsmobile. Leland's vision was to create a luxury car that would rival the best European models, and he succeeded in doing so with the launch of the first Cadillac in 1903. Here we explore the history of the Cadillac Motor Company, focusing on the role of Henry Leland along with Henry Ford in its initial launch and growth and the reasons behind its continued success over the years.

Henry M. Leland was born in Vermont in 1843 and grew up with a passion for mechanics and machinery. He trained as a machinist and worked in various industries, including firearms, bicycles and steam engines, before joining the emerging automotive industry in the late 19th century. In 1902, Leland and his son, Wilfred, were hired by the Henry Ford Company to appraise its assets and recommend a new course of action.

When Henry Leland was tasked with assessing the assets of the struggling Henry Ford Company, the company, much like the Detroit Automobile Company, was plagued by disputes between Ford and his investors. Leland was impressed by the company's machinery and production methods but found fault with its design and engineering. After evaluating the company, Leland advised its investors to invest more money and build a new car instead of liquidating the assets.

Unfortunately, Leland's advice was not heeded, and the investors decided to dissolve the Henry Ford Company. However, they recognized Leland's expertise and asked him to liquidate the assets and machinery. While overseeing the process, Leland noticed something remarkable—he saw immense potential in the 10 hp, single cylinder, engine he had designed for the company.

Leland believed that the engine had exceptional precision and quality, unlike anything else on the market at the time. Convinced that the engine could become the foundation for a remarkable automobile, he approached the investors and persuaded them to give him a chance to build a new car company around the engine. Leland was granted permission, and he named the new company after Antoine de la Mothe Cadillac, an exemplar of exploration and pioneering spirit. Cadillac displayed the new vehicles at the New

York Auto Show in January 1903, where the vehicles impressed the crowds enough to gather over 2,000 firm orders.

Leland's plan was to create a luxury car that would appeal to wealthy buyers who wanted something more than the standard utilitarian models that had dominated the market since its inception. He hired a team of skilled engineers and designers and set about designing a car that would be the epitome of style, comfort and performance. The first Cadillac, the Model A, was unveiled at the New York Auto Show in January 1903, and it was an instant sensation. The car's innovative features, such as electric lighting, a self-starter, and interchangeable parts, set it apart from its competitors and established Cadillac as an early leader in automotive innovation.

However, Leland's vision for Cadillac was not limited to luxury cars. He believed that the key to the company's success was precision engineering and quality manufacturing, which he demonstrated with his pioneering use of interchangeable parts and the establishment of a machine shop that was the envy of the industry. Leland's commitment to quality was so strong that he famously stated, "We don't build automobiles, we build Cadillacs." His insistence on precision and attention to detail paid off in the form of numerous awards and accolades, including the coveted Dewar Trophy, awarded by the Royal Automobile Club of England in 1908 for the first time to an American automaker.

Leland's legacy at Cadillac was not limited to his engineering prowess; he was also a shrewd businessman who understood the importance of branding and marketing. He hired the best designers and artists to create the Cadillac logo and advertising materials, which helped establish the brand's image as one of luxury, sophistication and exclusivity. Leland also recognized the importance of customer service and he established a network of dealerships and service centers that provided personalized attention to each customer. These efforts helped create a loyal customer base that was ready, willing and able to pay a premium for a Cadillac.

However, Leland's success at Cadillac was not without its challenges. In 1908, the company was acquired by General Motors, which was looking to expand its portfolio of brands. Leland initially resisted the sale, but he eventually agreed to stay on as president and continue to run the company as he saw fit. However, tensions between Leland and GM's management soon emerged, and in 1917, he was forced out of the company he had founded. Next, Leland went on to establish another famous car company, Lincoln, which was later acquired by Ford Motor Company.

Henry Ford's role in the history of Cadillac is significant, not only because of his involvement in the company's acquisition but also because of his influence on the automotive industry as a whole. Ford was a pioneer in mass production and assembly-line manufacturing, which revolutionized the way cars were made. He was also a fierce competitor who wanted to dominate the market, and he saw Cadillac as a threat to his ambitions. Ford's acquisition of Lincoln was a direct challenge to Cadillac's dominance in the luxury car market, and he poured resources into developing a car that would rival the Cadillac.

The rivalry between Cadillac Automobile Company and Lincoln Motor Company brands in the early days of their startup companies was intense and fierce. Both companies were founded in the early 1900s, Cadillac in 1902 and Lincoln in 1917.

At the time of the founding of Lincoln, Cadillac was already an established brand, having been in business for several years. Cadillac was known for its luxurious and high-performance vehicles, and it had a loyal customer base. Lincoln, on the other hand, was a newcomer to the industry in 1917 and had to work hard to establish itself as a serious contender in the luxury car market.

The rivalry between the two companies was fueled by their leaders' personalities and their desire to outdo each other. Cadillac and Lincoln constantly competed for market share, with each company trying to outdo the other in terms of performance, luxury, and innovation.

Despite the intense competition, both companies managed to thrive in the early years of their existence. Cadillac continued to innovate, introducing new technologies like the first mass-produced V16 engine, while Lincoln focused on creating luxurious and stylish vehicles that appealed to wealthy customers.

In the end, both companies left a lasting legacy in the automotive industry, with Cadillac becoming known for its luxury vehicles and Lincoln for its popular designs. The rivalry between the two companies may have been intense but the competition ultimately resulted in the creation of some of the most compelling cars in automotive history.

8

Buick

David Buick left an important, even indelible mark

David Buick

*"When I started Buick Motor Company, I didn't set out to make
a fortune. I just wanted to build a good car."*
—David Buick

David Buick was a Scottish-born inventor and entrepreneur who played a highly significant role in the early days of the automobile industry in the United States. He is best known for founding the Buick Motor Company, which would later become part of the startup and foundation of the now famous General Motors Corporation.

Buick was born in Arbroath, Scotland, in 1854 and immigrated to the United States with his family when he was just two years old.

The family settled in Detroit, Michigan. As a young man, Buick worked as a plumber and as a millwright, eventually starting his own business manufacturing plumbing supplies.

In the late 1800s, Buick became interested in the new technology of gasoline engines. He began experimenting with building his own engines and by 1899 Buick had developed a successful prototype gasoline-powered motor. Later, in 1902, with the help of a group of investors, Buick founded the Buick Auto-Vim and Power Company, with a goal of producing motor vehicles powered by his newly developed engine.

The early years of the Buick Motor Company were extremely difficult. Buick's first car, the Model B, was not successful and the company struggled to find its footing in the competitive automobile market. However, Buick's fortunes began to improve when it hired a young innovative engineer, named Walter L Marr, to develop another new engine design.

Marr's new design, known as the overhead-valve engine, was a breakthrough in automotive power source technology. It allowed for greater power and efficiency and was found to be much more reliable than earlier gasoline-powered engine designs. The first car powered with the overhead-valve engine, the Model D, was a success and sales of the Buick cars began to increase.

In 1908, David Buick sold his company to a group of investors, led by William C. (Billy) Durant, who would go on to found General Motors in September 1908. Buick remained with the company as a consultant, but he soon became disillusioned with Durant's management style and left the company the next year, 1909.

Despite his short tenure at the helm of the Buick Motor Company, David Buick's contributions to the automotive industry were significant. Buick's early experiments with gasoline engines helped lay the groundwork for the modern car and his company success with the overhead-valve engine helped establish Buick as a vitally important player in the still new automotive world.

The Buick Motor Company became one of the most successful and influential automakers of the early 20th century. Under the leadership of Durant and his successors at General Motors, Buick continued to innovate and produce groundbreaking new designs, including the first mass-produced four-wheel-drive vehicle and the first V-8 designed engine.

Another important aspect of the Buick legacy was the role that the Howe Buick division of General Motors played in the creation of other automotive innovators. Howe Buick was a franchised dealership in Detroit which specialized in customizing and modifying V-8 powered cars for racing and other high-performance applications.

Among the employees of Howe Buick were a number of young engineers and mechanics who would go on to become major figures in the still young automotive industry. One of those was Walter P. Chrysler, who founded the Chrysler Corporation.

Buick has been one of the most recognized brands in the automobile industry, known for its luxury and performance vehicles. The post-war years of Buick were marked by significant changes in the auto industry and the company responded by introducing new models that reflected the changing tastes of consumers.

After World War II, Buick shifted its focus to producing larger, more luxurious cars to meet the growing demand for high-end vehicles. In 1948, Buick introduced the Roadmaster, which quickly became one of the most popular models in the company's history. The Roadmaster was a large, spacious vehicle that was designed for comfort and luxury, with features like power windows, power seats and air conditioning.

In the 1950s, Buick continued to innovate with new models like the Skylark, which was introduced in 1953 and featured a sleek, sporty design that was ahead of its time. The Skylark was a convertible that was designed to appeal to younger buyers who wanted a car that was both stylish and fun to drive.

In the 1960s, Buick continued to produce large, luxurious cars, but also began to introduce more performance-oriented models like the Gran Sport. The Gran Sport was a high-performance version of the company's popular Skylark model and featured a powerful V-8 engine and sport-tuned suspension.

Today, Buick continues to produce a range of high-quality vehicles that are designed to meet the needs of modern consumers. Some of the most popular models currently offered by Buick include the Encore, Envision and Enclave, small to medium-sized luxury SUVs that offer spacious seating and advanced safety features.

9

Ford

#1 Auto industry pioneer:
The legendary Henry Ford

Henry Ford, in his Model T

*"If I would have asked people what they wanted,
they would have said faster horses."*
—Henry Ford

Throughout modern history, Henry Ford has largely been credited with launching the US auto industry. Born in 1863 in Springwells Township, Michigan, Ford grew up on a farm and demonstrated early interest in mechanics and technology. In 1879, Ford left home to work as an apprentice machinist in Detroit, first with James F. Flower & Bros., and later with the Detroit Dry Dock Co. In 1882, he returned to Dearborn to work on the family farm, where he became adept at operating the Westinghouse portable steam engine.

Ford worked as an apprentice machinist and then as an engineer for several companies, including the Henry Leland Company, which later became the Cadillac Motor Company. Ford was later asked by investors to run the Henry Leland Company which he did for a short time, though he was not happy with its direction and left to start his own auto manufacturing company.

Henry Ford started Ford Motor Company in 1903. Within just a few years, Ford was producing cars that were both affordable for the average American and easy to operate, thanks to his innovative manufacturing techniques and emphasis on efficiency.

Ford's decision to focus on gas-powered cars rather than electric ones was what became a crucial factor in the success of his company in the broader auto industry. At the turn of the 20th century, both gas and electric cars were being developed and marketed. Electric cars were popular among wealthy city dwellers and appreciated for their quiet operation and lack of pollution. However, they were expensive, and had limited range, which made them impractical for most people. In essence, the same challenges existed for electric vehicles when Henry Ford was starting out in the early 1900s that still exist today. More than a century after Henry Ford chose gas-powered cars over battery electric, the country and the world are still trying to overcome problems with the rollout of electric vehicles.

Gas-powered cars on the other hand, were cheaper to produce and could travel longer distances. They were also louder and emitted more pollution, but those drawbacks were outweighed by two important advantages: practicality and affordability. Henry Ford recognized that the key to creating a successful automobile industry was to produce cars the average person could more easily afford to purchase and gas-powered cars were the way to do that.

Ford's decision to focus on gas-powered cars was not without controversy, however. His close personal friend, Thomas Edison, inventor of the electric light bulb and many other devices, was a strong advocate for electric cars. Thomas Edison believed that electric cars were cleaner and more efficient than gas-powered cars and Edison even designed a battery he claimed would allow electric cars to travel long distances. Edison and his second wife, Mina, were close friends with Henry Ford and his wife, Clara, and the Edisons often tried to persuade Ford to switch to electric cars.

Despite their friendship and Edison's persuasive argument, Ford remained committed to gas-powered cars. Ford believed that they were the best choice for the average person, and he was unwilling to compromise on his vision for mass-produced affordable automobiles. That decision proved to be the right one at the time, as gas-powered cars quickly became the dominant form of transportation in the United States and around the world.

Ford's tremendous success was due in large part to his innovative manufacturing techniques, which allowed him to produce cars quickly and efficiently. Ford was an early developer of the assembly line process and a method of production in which each worker performs a specific task along a moving conveyor belt. This greatly increased production speeds and reduced costs making it possible to produce cars on a much larger, more efficient scale. Ford also introduced the concept of interchangeable parts, which allowed parts to be mass produced and then assembled into different cars as needed. This made repairs and maintenance easier and less expensive, further increasing the appeal of his cars to the average consumer.

It is important to note that though Henry Ford gets much credit for the assembly line process, Ransom Eli Olds was actually utilizing an earlier version of the assembly line concept a full 10 years before Henry Ford when he produced the Curved Dash Oldsmobile in

Lansing, Michigan. The primary difference between Olds' version of the assembly line process and that of Henry Ford, was that Olds' assembly line was configured where the car remained in one place and autoworker assemblers moved down the assembly line. Henry Ford's version was the opposite. In Ford's version of the assembly line, the workers remained stationary, and the cars advanced down the assembly line. Henry Ford's version won out over time. Today, virtually all major automakers around the world utilize Henry Ford's version of the assembly line and do so very successfully. The Ford version of the assembly line won the day and the industry.

In addition to the generations in manufacturing, Ford was also a pioneer in marketing and advertising. He recognized the importance of creating a brand image and promoting his cars as symbols of American ingenuity and progress. Ford sponsored race cars and other events to demonstrate the speed and durability of his cars and he used slogans like "Ford, the universal car," to emphasize their appeal to people of all classes and backgrounds.

Like Ransom Eli Olds before him, Henry Ford knew he had his hands full designing and building cars, figuring out the power source technology for propulsion and accessing the materials and parts to build them. Both Olds and Ford knew they needed to offload the distribution processes for their rolling machines, so both were early pioneers in the development of the franchised dealer network. While Ransom Eli Olds had franchised dealers as early as 1898, the first contracted franchised dealer with Ford dates to 1908, the same year as the rollout of the Model T.

Ford's success was not without controversary. He was known for his staunch opposition to labor unions, and his use of somewhat violent tactics to suppress worker protest. He also faced criticism due to his antisemitic views and his admiration for Nazi Germany. Despite those personal flaws, however, Ford's impact on the auto industry and therefore on American society, as a whole, cannot be denied.

Ford was not only passionate about revolutionizing the automobile industry but also had a keen interest in private aviation and flying. Additionally, he had a personal involvement in designing, building, and selling bicycles. These two seemingly unrelated pursuits reveal Ford's relentless curiosity, thirst for innovation, and his drive to improve transportation in all forms.

Ford's fascination with aviation was sparked in the early 1900s when the Wright brothers successfully achieved powered flight. Inspired by their accomplishment, he began investing in and experimenting with aviation technologies.

In 1909, Ford formed the Ford Aeronautical Company, aiming to develop his own aircraft. Although this venture was short-lived, it demonstrated his enthusiasm for aviation and his desire to be at the forefront of technological advancements. Furthermore, Ford's personal interest in flying extended beyond mere experimentation. In the 1920s, he established the Ford Air Transport Service (FATS), an airline that provided passenger and mail services. FATS played a crucial role in transporting Ford's executives and employees across the country, ensuring efficient communication and coordination within his vast industrial empire.

Ford's involvement in designing, building, and selling bicycles may seem inconsequential compared to his achievements in the automobile industry. However, it is a testament to his entrepreneurial spirit and determination to create affordable transportation options for the masses. In his youth, Ford worked as a bicycle mechanic, and this experience instilled in him a deep understanding of mechanical engineering and the principles of transportation.

Henry Ford had a professional relationship with early aviation flight pioneers at Kitty Hawk, Orville and Wilbur Wright, though they were not close friends. Ford greatly admired the Wright brothers and their achievements in aviation. In fact, he even purchased one of their airplanes, called the Wright Model B, in 1910.

As for Henry Ford's personal involvement in aviation, although he was not an active private pilot himself, he had a strong interest in aviation and owned several aircraft, he did not frequently fly them personally. He preferred to leave the piloting to professionals.

In 1896, Ford built his first experimental automobile, the Quadricycle, which utilized bicycle parts. This early creation showcased his ability to bridge the gap between bicycles and automobiles, laying the foundation for his future groundbreaking achievements. Ford's knowledge of bicycles also influenced his approach to automobile design, as he aimed to create vehicles that were reliable, lightweight, and affordable for the average person, much like bicycles.

Moreover, Ford's involvement in the bicycle industry was not limited to his early experiments. When he established the Ford Motor Company, in 1903, one of his other major business ventures was acquiring the Detroit Bicycle Company. This acquisition allowed him to tap into the existing bicycle market and leverage its distribution networks to promote his new automobile venture.

Henry Ford's personal interest in private aviation and flying, as well as his involvement in designing, building, and selling bicycles, highlights his unwavering commitment to transportation innovation. These pursuits were not mere hobbies but rather reflections of his relentless drive to improve and transform the way people traveled. Ford's contributions to aviation and his understanding of bicycles as a precursor to his automotive success demonstrate his visionary mindset and his ability to identify opportunities for advancement in various modes of transportation.

Ford Motor Company has been important to the auto industry through the ages having experienced significant growth and success from the World War II time period to present day. During the 1950s and 1960s, Ford became a dominant force in the automotive industry with the introduction of iconic models such as the Ford Thunderbird and the Ford Mustang. These models were popular among consumers due to their sleek design and high-performance capabilities.

In the 1970s, Ford Motor Company continued to innovate with the introduction of the Ford Pinto, which was the first meaningful American-made subcompact car. However, the company faced significant challenges during this time period due to the oil crisis. That led to a decline in the sales of large cars and an increase in demand for smaller, more fuel-efficient vehicles.

In the 1980s and 1990s, Ford continued to expand its product line with the introduction of the Ford Taurus and the Ford Explorer, which became two of the company's most successful models. The Taurus was popular among consumers due to its innovative design and safety features, while the Explorer became a popular choice for families due to its spacious interior and off-road capabilities.

In the 2000s, Ford faced significant challenges due to the global financial crisis, which led to a decline in sales and profits. However, the company was able to recover by focusing on innovation and technology, with the introduction of models such as the Ford Fusion and the Ford Escape. Ford was the only major US-based automaker that was able to survive the recession of 2008-09 without taking bridge loans as a governmental handout. The others needed those and accessed them. That distinction has set Ford apart from the other major domestic automakers.

Today, Ford has proven to be a leader in the automotive industry with a wide range of popular models, including the Ford F-150, the Ford Mustang and the Ford Explorer. The F-150 is the best-selling vehicle in the United States and has been a staple of the company's product line for over 40 years.

Ford Motor Company's growth and success over the years can be attributed to its ability to innovate and adapt to changing consumer demands. The company has produced many popular cars over time, including the Ford Thunderbird, Mustang, Pinto, Taurus, Explorer, Fusion, Escape, F-150, and more. The Ford F-150 has held the Number One position for sales of any model for over

45 years. And Ford was one of first traditional automakers in the world to move toward mass-produced electrification with the rollout in 2021 of the all-electric Ford Mustang Mach E and in 2022 with the all-electric Ford F-150 Lightning.

10

Rolls-Royce

British luxury led early with Rolls Royce

Charles Rolls and Henry Royce

"The quality will remain long after the price is forgotten."
—Charles Rolls

"I do not believe in the word impossible."
—Henry Royce

Rolls-Royce is a British luxury car company founded in 1904 by Charles Rolls and Henry Royce. The company quickly became known for its high-quality, handcrafted vehicles renowned for their exceptional engineering and craftsmanship.

In the early years, Rolls-Royce produced luxury cars that were popular with wealthy individuals and business leaders. The company also developed aircraft engines, which played a crucial role in World War I and helped establish Rolls-Royce as a leading manufacturer of aviation engines.

Over the years, Rolls-Royce continued to innovate and expand its product line. The company produced a range of luxury cars, including the iconic Silver Ghost and the Phantom, both of which became symbols of wealth and prestige around the world.

In addition to its success in the automotive industry, Rolls-Royce established itself as a leading manufacturer of aircraft engines. The company's engines have powered some of the most iconic planes in history, including the Spitfire, the Lancaster Bomber and the Concorde.

Famous movies and television shows that have featured Rolls Royce cars include:

- *The Thomas Crown Affair* (1968 and 1999)
- *Entourage* (2004-2011)
- *Downton Abbey* (2010-2015)
- *The Great Gatsby* (2013)
- James Bond films (various)

Famous people who have owned Rolls-Royce cars include:

- Queen Elizabeth II
- John Lennon
- Elvis Presley
- Jay Leno
- David Beckham
- Simon Cowell
- Beyoncé
- Kim Kardashian
- Taylor Swift

Today, Rolls-Royce continues to be a leading manufacturer of luxury cars and aircraft engines. The company's vehicles are still renowned for their exceptional quality and craftsmanship, and are favored by celebrities, business leaders and royalty around the world.

Rolls-Royce's success and growth over the years can be attributed to its unwavering commitment to quality, innovation and craftsmanship. Charles Rolls and Henry Royce's vision for a luxury car company that would set new standards for engineering and design has been realized in the enduring, longterm, legacy of Rolls-Royce.

11

Bugatti

The French blessed the world with Bugatti

"Nothing is too beautiful, nothing is too expensive."
—*Ettore Bugatti*

"The Bugatti is the world's most famous automobile."
—*Ralph Lauren*

The Bugatti car company is a French luxury car manufacturer that has been around since 1909. It was started by Ettore Bugatti, an Italian-born engineer and designer who had a passion for creating high-performance automobiles. The company was initially based in Molsheim, a small town in the Alsace region of France.

Bugatti quickly gained a reputation for producing some of the most innovative and stylish cars of the era. Its early models were known for their sleek lines, powerful engines and exceptional handling. Bugatti's cars were also popular among racing enthusiasts, and the company won several prestigious races, including the Targa Florio and the Grand Prix de l'ACF.

Over the years, ownership of the Bugatti car company has changed hands several times. In the 1950s, the company was sold to the Hispano-Suiza Group, and later to the French car manufacturer, Renault. In the 1980s, Bugatti was purchased by the Italian entrepreneur, Romano Artioli, who set out to revive the brand with a new line of high-performance sports cars.

Despite Artioli's efforts, the Bugatti brand struggled financially and was eventually acquired by Volkswagen Group in 1998. Under Volkswagen's ownership, Bugatti has continued to produce some of the most exclusive and expensive cars in the world, including the Veyron and the Chiron. These cars are known for their exceptional performance, luxurious features and unique design elements.

Movies and TV shows featuring Bugatti cars:

- *The Great Gatsby* (2013): A Bugatti Type 57 is featured in the opening scene.
- *The Thomas Crown Affair* (1999): A Bugatti Type 41 Royale is stolen in this film.
- *Top Gear* (2002-2015): Bugatti Veyron and Chiron have been reviewed and tested on this popular car show.

Famous people who own Bugatti cars:

- Cristiano Ronaldo: The soccer star owns a Bugatti Veyron.
- Floyd Mayweather, Jr.: The boxer owns a Bugatti Veyron and a Bugatti Chiron.

- Jay Leno: The talk show host and car enthusiast owns a Bugatti Type 57 Atlantic and a Bugatti Veyron.
- King Leopold III of Belgium: Owned a Bugatti Type 44
- Roland Garros: French aviator and tennis player who owned a Bugatti Type 35
- Prince Rainier III of Monaco: Owned a Bugatti Type 57C
- Ralph Lauren: Fashion designer owns a Bugatti Type 57SC Atlantic
- Simon Cowell: TV personality owns a Bugatti Veyron
- Tom Brady: NFL quarterback owns a Bugatti Veyron

Today, the Bugatti car company is still based in Molsheim, France, and continues to be a symbol of luxury and innovation in the automotive industry. Its cars are highly sought after by collectors and enthusiasts around the world, and the brand remains an icon of French engineering and design.

12

Alfa Romeo

From Italy with love

Alexandre Darracq

*"Alexander Darracq was a visionary and his legacy
lives in the Alfa Romeo."*
—Unknown

The Alfa Romeo car company is an Italian automobile manufacturer in existence since 1910. It was founded in Milan, Italy, by a group of investors, including Alexandre Darracq, who was a French car maker. The company was initially called Società Anonima Italiana Darracq (SAID), but it was renamed Alfa Romeo in 1915.

Alfa Romeo has a rich history in the automotive industry, producing some of the world's most iconic cars. The company has been involved in motorsport since its inception, which has helped to establish its reputation for producing high-performance vehicles.

The brand has won numerous championships in various racing categories over the years, including Formula One, touring car racing and sports car racing.

Over the years, Alfa Romeo has undergone several ownership changes. In 1915, the company was taken over by Nicola Romeo, an Italian entrepreneur, which led to the renaming of the company to Alfa Romeo. During World War II, the company was taken over by the Italian government and used to produce military equipment.

In 1986, the company was acquired by the Fiat Group, which led to a significant increase in production and sales. In 2007, the Alfa Romeo brand was merged with Fiat's other brands, including Lancia and Fiat itself, to form Fiat Chrysler Automobiles (FCA). In 2021, FCA merged with French automaker Groupe PSA to form Stellantis, which is the world's fourth-largest automaker. In 2014, FCA announced plans to spin off Alfa Romeo as a separate entity, but this did not happen.

Famous television shows and movies that have featured Alfa Romeo cars include:

- *The Graduate* (1967) - Alfa Romeo Spider
- *The Godfather* (1972) - Alfa Romeo Giulia
- *Quantum of Solace* (2008) - Alfa Romeo 159
- *John Wick: Chapter 2* (2017) - Alfa Romeo Giulia Quadrifoglio
- *Killing Eve* (2018) - Alfa Romeo Giulia

Famous people who have owned Alfa Romeo vehicles:

- Enzo Ferrari
- Juan Manuel Fangio
- Rudolf Caracciola
- Ugo Sivocci
- Nino Farina

- Carlo Chiti
- Giuseppe Campari
- Tazio Nuvolari
- Vittorio Jano
- Ken Miles

Alfa Romeo is a historic and iconic Italian car manufacturer that has been in existence for over a century. Even though the company has undergone several ownership changes over the years it has maintained its reputation for producing high-performance vehicles. Alfa Romeo continues to be a popular brand among car enthusiasts worldwide.

13

Audi

Germany adds Audi to list of high-impact car makers

August Horch

"Audi is a brand with strong heritage and a great reputation for building some of the finest cars in the world."
—Unknown

Audi, a German car company, has a long and rich history dating back to the early 20th century. The company was founded by August Horch, a German engineer who had previously worked for both Benz and Daimler. Horch was a pioneer in the automotive industry and was instrumental in the development of some of the earliest European cars.

In 1899, Horch founded his first car company, which he named after himself, the "A. Horch & Cie." The company quickly gained a reputation for producing high-quality cars that were both reliable and durable. However, after a few years, Horch experienced disagreements with his business partners and was forced to leave the company he founded.

Undeterred, Horch decided to start a new car company. Unable to use his own name as it was already trademarked by his former company, he decided to use the Latin translation of his name, which is "audi." Thus, the Audi car company was born.

The early years of Audi were marked by rapid growth and expansion. In 1909, the company introduced its first six-cylinder car, which was considered a technological marvel at the time. Audi's six-cylinder car was fully capable of reaching speeds of up to 80 km/h, which was a remarkable achievement for the era.

In the years that followed, Audi continued to innovate and develop new technologies. In 1911, the company introduced its first four-wheel drive car, a major breakthrough in automotive engineering. The car was designed for use in rugged terrain and was highly popular with farmers and other rural workers.

During the First World War, Audi was heavily involved in the production of military vehicles. The company produced a wide range of vehicles for the German army, including trucks, ambulances and even tanks. This experience helped the company to develop its engineering capabilities and cemented its reputation as a leading automotive manufacturer.

After the war, Audi resumed its focus on producing high-quality passenger cars. In the 1920s, the company introduced a number of new models, including the Audi Type K, a popular luxury car. The car was designed to compete with other luxury brands of that era such as Mercedes-Benz and Rolls-Royce and was highly successful.

The Great Depression of the 1930s had a significant impact on the automotive industry. Sales of luxury cars declined sharply, and many car companies were forced to close their doors. Audi struggled to stay afloat during this difficult period.

In 1932, Audi merged with three other German car companies to form Auto Union. The new company was a major force in the automotive industry and produced a wide range of vehicles, including luxury cars, sports cars, and even race cars. The company's racing team was highly successful and won numerous championships during the 1930s.

The outbreak of World War ll in 1939 had a major impact on the automotive industry. Auto Union was forced to stop producing passenger cars and instead focused on producing military vehicles for the German army. The company produced a wide range of vehicles, including tanks, trucks and even aircraft engines.

After the war, due to military devastation, Germany was in ruins and the automotive industry was in shambles. Auto Union disbanded, and its factories were seized by the Allies. However, the Audi brand name was able to live on. In the 1950s, the company resumed production of passenger cars.

The post-war years were marked by significant innovation and development in the auto industry. Audi was at the forefront of this revolution and introduced a number of new technologies, including the first fully automatic transmission. The company also developed a number of new models, including the Audi 100, which was a highly popular midsize car.

In the 1960s, Audi introduced a new range of sporty cars, including the Audi 60 and Audi 75. These cars were designed to appeal to a younger, more affluent demographic and were highly successful. The company also continued to innovate and develop new technologies, including the first electronic fuel injection system. In 1964, Volkswagen bought 50 percent of Audi and 18 months later bought the remainder of the company rolling it into its larger Volkswagen Automotive Group.

In the 1970s, Audi faced a major challenge when it was accused of producing cars that were prone to sudden unintended acceleration. The controversy was a major blow to the company's reputation and led to a significant decline in sales, though Audi was able to recover from the scandal and continued to produce high-quality cars.

Audi has been one of the most successful automobile manufacturers in the world. The company has managed to grow and expand its business significantly since 1990, establishing itself as one of the top three luxury car brands globally.

While still a relatively small player in the luxury car market, the company had a clear vision of where it wanted to be and what it needed to do to get there. Audi focused on producing high-quality cars that were technologically advanced and had a unique design. The company invested heavily in research and development, which helped it to develop new technologies and innovative designs. This focus on innovation and quality helped Audi to establish itself as a premium brand.

In the early 2000s, Audi expanded its product line introducing the A3, A4, A5 and A6 models, which helped to increase its market share and attract new customers. These models were designed to appeal to a wider range of customers, including younger buyers who were looking for a more affordable luxury car. Additionally, Audi introduced the Q7, which was a huge success and helped the company to enter the SUV market.

In recent years, Audi has continued to grow and expand its business. The company has focused on developing new technologies and improving the performance of its cars. Audi has also expanded its product line to include electric and hybrid vehicles. The company's e-tron SUV has been a huge success, and Audi plans to introduce more electric and hybrid models in the coming years.

Audi has also focused on expanding its global presence. The company has established manufacturing facilities in China and Mexico, which has helped it to reduce production costs and improve efficiency. Audi has also increased its marketing efforts in emerging markets, such as China and India, which has helped it to attract new customers and increase sales.

Did you know that many past or present political leaders around the globe have found favor with Audi cars for their personal or governmental use? These include:

- Former German Chancellor Angela Merkel
- Russian President Vladimir Putin
- Indian Prime Minister Narendra Modi
- Chinese President Xi Jinping
- Former Japanese Prime Minister Shinzo Abe
- South Korean President Moon Jae-in
- French President Emmanuel Macron
- Former British Prime Minister Theresa May
- Spanish Prime Minister Pedro Sánchez
- Canadian Prime Minister Justin Trudeau

Audi has been able to achieve significant growth in recent years. The company's focus on innovation, quality, and expanding its product line has helped it to maintain itself as a premium brand in the luxury car market. Audi's expansion into new markets and its focus on developing new technologies and electric vehicles has helped it to stay ahead of its competitors. With its continued focus on innovation and quality, Audi is well-positioned to continue its growth and success.

14

GMC

From the earliest years, GMC is proudly GM's truck brand

Billy Durant, GM president

"GMC trucks are built to handle the toughest jobs with ease, delivering the reliability and durability you expect from a professional grade truck."
—*Unknown*

General Motors (GM) is one of the largest automobile manufacturers in the world. The company has a long and rich history dating back to the early 1900s (see other chapters on General Motors brands). One of the most significant divisions of GM is the GMC Trucks division, founded in 1911. GMC Trucks is known for producing high-quality, durable and reliable trucks used for both commercial and personal purposes.

The history of GMC Trucks can be traced back to the Rapid Motor Vehicle Company, which was founded in 1901 by Max Grabowsky. The company was in Pontiac, Michigan, and produced trucks and other commercial vehicles. In 1909, one year after the start of General Motors, Grabowsky sold the Rapid Motor Vehicle Company to GM, which was looking to expand its portfolio of automobile brands.

After the acquisition of the Rapid Motor Vehicle Company, General Motors decided to merge the company with another truck manufacturer, the Reliance Motor Car Company. The two companies were merged to form the General Motors Truck Company in 1911. The new company was tasked with producing trucks for both commercial and personal use.

The early years of the General Motors Truck Company were marked by significant growth and expansion. The company quickly gained a reputation for producing high-quality trucks that were both durable and reliable. The trucks were used by businesses and individuals all over the country, and the company became one of the largest volume truck manufacturers in the United States.

GMC truck climbing Pike's Peak in 1909

The true story of the 1909 GMC truck climbing Pike's Peak is a remarkable feat of automotive history. In 1909, a team of engineers and drivers from the newly formed GM embarked on a daring challenge to prove the capabilities of their vehicles. They aimed to climb Pike's Peak, one of the most treacherous and challenging mountain roads in the United States.

Pike's Peak, located in Colorado, rises to an elevation of 14,115 feet (4,302 meters). Its steep grades, rugged terrain, and thin air made it an arduous task for any vehicle, let alone a truck from the early 1900s. However, GMC engineers were determined to demonstrate the power and reliability of their machines.

The chosen vehicle for this ambitious endeavor was a 1909 GMC Model 60 truck, equipped with a four-cylinder engine producing 30 horsepower. The truck featured innovative engineering and design elements for the time, including a sturdy frame, advanced cooling system, and robust brakes.

On August 11, 1909, the GMC team began their ascent up Pike's Peak. The journey was fraught with challenges right from the start. The unpaved road was riddled with rocks, loose gravel, and steep inclines. As the truck climbed higher, the thin air caused the engine to lose power, making the ascent even more difficult.

Undeterred, the team pressed on, employing every ounce of skill and determination. They frequently stopped to cool the engine, allowing it to regain its strength. The crew worked tirelessly, making adjustments and overcoming various obstacles along the way.

Finally, after hours of intense effort, the GMC truck reached the summit of Pike's Peak. It was an extraordinary achievement, demonstrating the truck's exceptional durability and performance in extreme conditions. The feat made headlines across the country and showcased the capabilities of the GMC truck.

Evolution of the brand name

In 1916, leaders at GM rebranded the General Motors Truck Company as the GMC Truck division. The new name was more in line with the company's focus on producing high-quality trucks designed for commercial and personal use. The GMC Truck division became one of the most recognizable brands in the trucking industry.

Also in 1916, William Warwick drove a GMC truck carrying a ton of Carnation® canned milk from Seattle, Washington to New York City, New York and back. It took 21 weeks. Fifty years later GMC recreated the promotion. The 1966 trip took only six days.

Cannonball Baker breaks in GMC

In 1927, a daredevil race driver, Erwin George "Cannonball" Baker, broke the cross-country record driving a GMC truck from New York City to San Francisco in 5 days and 17.5 hours. The legendary driver was an accomplished motorcycle and automobile racer who became widely known for his long-distance driving abilities and endurance.

Baker pushed himself to the limit, driving almost non-stop across the diverse landscapes of the United States. He navigated through various road conditions, ranging from smooth highways to rugged terrain and challenging weather. The route he took passed through states including Pennsylvania, Ohio, Indiana, Illinois, Missouri, Kansas, Colorado, Utah, Nevada, and finally, California.

World War I years

During World War I, GMC Trucks played a critical role in the war effort. The company produced a variety of trucks and other vehicles used by the military. The trucks transported troops, supplies and equipment to the front lines. The success of the company during the war helped to solidify GMC Trucks' reputation as a reliable and dependable truck manufacturer.

Following World War I, GMC Trucks continued to grow and expand. The company introduced a variety of new models, including the T-series, designed for heavy-duty commercial use. The T-series was a huge success, and it helped to cement GMC Trucks' position as one of the leading truck manufacturers in the world. During the 1920s and 1930s, GMC Trucks continued to innovate and introduce new models.

During World War II, GMC Trucks again produced vehicles for the military. The trucks were used to transport troops, supplies and equipment to the front lines.

After the end of World War II, GMC Trucks embraced the Hydra-Matic transmission, an automatic transmission that was more efficient and reliable than previous transmissions. The Hydra-Matic transmission became a popular choice for businesses and individuals who needed a powerful and reliable truck for heavy-duty work.

During the 1950s and 1960s, GMC Trucks continued to grow and expand. The company introduced a variety of new models, including the C-series, which was designed for heavy-duty commercial use. The C-series success helped strengthen GMC Trucks as a leading truck manufacturer in the world.

During the 1970s and 1980s, GMC Trucks introduced the Sierra, a full-size pickup truck designed for personal use. The Sierra became a popular choice for individuals needing a powerful and reliable truck.

Between 2000 and the present, GMC has undergone significant growth and product development. GMC has established itself as a leading brand in the pickup truck and SUV market, with a focus on premium features and design.

One of the key drivers of GMC's growth in recent years has been the popularity of its Sierra pickup truck lineup. The Sierra has consistently been one of the top-selling full-size pickups in the US, thanks in part to its rugged design and advanced technology features. GMC has also expanded its lineup of SUVs, with the introduction of the Acadia, Terrain and Yukon models.

I'm proud to say that I have been the appreciative owner of not one, but two, GMC Terrains. They were both a pleasure to drive and to own. They carry a distinctive look and are enjoyable to drive with quick responsiveness, latest and greatest creature comforts and driving technology. For those in the small SUV market, I am quite honored to recommend the Terrain.

In terms of product development, GMC has focused on incorporating advanced technology features into its vehicles. The Sierra, for example, includes a head-up display, adaptive cruise control and a rearview camera system. The Yukon SUV offers a range of advanced safety features, including forward collision alert, lane departure warning and automatic emergency braking.

Another area of focus for GMC has been overall vehicular design. The company has emphasized a bold, distinctive design aesthetic across its lineup, with features such as signature LED lighting and premium materials. This focus on design has helped to differentiate GMC from other brands in the highly competitive pickup truck and SUV market.

GMC is likely to continue its focus on developing premium pickup trucks and SUVs, while also incorporating advanced technology features and design elements. As the market for electric and hybrid vehicles continues to grow, it is possible that GMC may also explore options for developing green vehicles that align with its brand identity. Overall, the GMC division of GM has demonstrated strong growth and product development in recent years, positioning itself as a leading player in the pickup truck and SUV market.

15

Chevrolet

See the USA in a Chevrolet

Chevrolet brothers: Gaston, Arthur and Louis

The Chevrolet brand has a rich history with its founders,
Louis Chevrolet and William C Durant,
creating a legacy that continues today."
—Car and Driver

Chevrolet is one of the most recognizable and even iconic American automotive brands. The company has a long and storied history, dating back over 100 years. From its humble beginnings as a small car manufacturer in Detroit, Michigan, to its current status as one of the world's largest automotive companies, Chevrolet has played a significant role in shaping the automotive industry.

The early years

The story of Chevrolet Motor Company began in 1911, when Swiss race car driver Louis Chevrolet and his brothers Arthur and Gaston founded the Chevrolet Motor Company in Detroit. The Chevrolet brothers had previously worked for the Buick Motor Company, though they left the company to pursue their own venture.

In the early 20th century, Louis Chevrolet established himself as a skilled race car driver. In 1905, he participated in the inaugural Vanderbilt Cup, one of the most prestigious automobile races of the time held in Long Island, New York. The Vanderbilt Cup attracted the best drivers from around the world, and winning it was a testament to one's skill and determination.

During the race, Chevrolet found himself competing against some of the most renowned drivers of the era, including William K. Vanderbilt himself, the founder of the Vanderbilt Cup. Chevrolet was driving a powerful but relatively unknown car, the Fiat, and despite being a newcomer, he showed exceptional talent and fearlessness on the track.

As the race progressed, Chevrolet's driving skills and the performance of his car caught the attention of the spectators and fellow racers. He quickly emerged as a serious contender, challenging the established champions and setting a blistering pace. Despite encountering mechanical issues during the race, Chevrolet showcased his ingenuity by making repairs on the go and swiftly getting back on track.

In a stunning turn of events, Louis Chevrolet, the underdog in the race, surpassed all expectations and triumphed over his competitors, crossing the finish line in first place. His victory not only marked a significant milestone in his racing career but also solidified his reputation as a brilliant driver and engineer.

This remarkable achievement propelled Chevrolet into the limelight and earned him a well-deserved reputation in the racing community. His success on the track eventually led him to co-found the Chevrolet Motor Car Company in 1911 with the backing of influential businessmen, including William C. Durant.

From Chevrolet Brothers to William C. (Billy) Durant

Chevrolet Motor Company initially produced a range of high-performance cars, but the company struggled to gain traction in the highly competitive automotive market. In 1916, the Chevrolet brothers sold their company to General Motors, which was then headed by William C. (Billy) Durant.

Durant had previously founded General Motors in 1908 and had built the company into a major force in the automotive industry. The acquisition of Chevrolet Motor Company added a new brand to the General Motors portfolio and allowed the company to expand into new markets.

Under General Motors' ownership, Chevrolet began to focus on producing affordable, mass-market cars. The company's first major success came in 1923, with the introduction of the Chevrolet Superior, which was a popular midsize car that was both affordable and reliable.

The Great Depression

The 1930s were a difficult time for the automotive industry, as the Great Depression gripped the country. Many car manufacturers were forced to close their doors, but Chevrolet managed to weather the storm, thanks to its reputation for producing affordable, reliable cars.

During this time, Chevrolet introduced several new models, including the Master and Deluxe, both popular with consumers. The company also introduced the first car radio, which was a major innovation at the time.

World War II

During World War II, Chevrolet played a critical role in the United States' war effort. The company produced a broad range of military vehicles, including trucks, utility vehicles and tanks. Chevrolet also manufactured aircraft engines for the war effort.

After the war, Chevrolet resumed production of its popular cars. The company introduced several new models, including the Bel Air, which was a popular midsize car. The Bel Air was known for its stylish design and powerful engine, and it quickly became one of Chevrolet's most popular models.

The 1950s and 1960s

The "See the USA in Your Chevrolet" ad campaign was launched by Chevrolet in the 1950s. The campaign was aimed at promoting the idea of road trips and encouraging people to explore the country in their Chevrolet cars. The campaign featured catchy jingles and slogans that became ingrained in popular culture. The campaign was successful in creating a sense of adventure and freedom associated with owning a Chevrolet. The campaign ran for several decades and is still remembered as one of the most iconic ad campaigns in American history.

More broadly, the 1950s and 1960s were a time of great change and innovation for the automotive industry, and Chevrolet was at the forefront of many of these changes. The company introduced several new models during this time, including the Corvette, which was a two-seater sports car that quickly became a symbol of American automotive culture.

Chevrolet also introduced the Impala, which was a popular full-size car that was known for its stylish design and carried a powerful engine. The Impala quickly became one of Chevrolet's most popular models, and it remains a popular collector car today. The iconic 1957 Chevy was an early Impala model design.

I was born in 1957 and my parents bought a new 1957 Chevrolet that year. My mom's youngest brother asked to borrow the car for his senior year prom at Breckenridge, Missouri High School. On the way to the prom, a driver coming from the other direction crossed the center line and hit my uncle and our '57 Chevy head-on. My uncle and his prom date were killed instantly. Passengers in the back seat (friends and fellow prom participants) survived the accident, fortunately. I remember the '57 Chevy fondly for the institutional icon it became. And I remember it sorrowfully for the accident that killed my uncle Bobby Merrifield.

In the 1960s, Chevrolet also introduced several new technologies, including fuel injection and turbocharging, the first of its type for the Chevrolet brand. These technologies helped to improve the performance and operational efficiency of Chevrolet's cars, and they helped to establish the company as a leader in automotive innovation.

The 1970s and 1980s

The 1970s and 1980s were a time of great change for the automotive industry, as concerns about fuel efficiency and emissions led to the development of new technologies. During this time, Chevrolet introduced the Camaro and the Chevette.

The Camaro was a popular sports car that became well-known for its powerful engine and stylish design. The Chevette was a small, fuel-efficient car that was designed to appeal to consumers who were concerned about rising fuel costs.

In the 1980s, Chevrolet introduced several new technologies, including the use of computer-controlled fuel injection and electronic ignition systems. These technologies helped to improve the performance and efficiency of Chevrolet's cars.

The modern era

In the 1990s and 2000s, Chevrolet continued to innovate and introduce new models. The company introduced several new SUVs and trucks, including the Silverado pickup (1999); Volt, an early plug-in hybrid car (2011); and the Chevrolet Bolt, an affordable all-electric vehicle (2017). GM announced in 2023 it would discontinue the Bolt to make room for several new all-electric vehicles planned for their lineup.

The iconic Chevrolet brand has made it into many famous songs. Those include:

- "409" by The Beach Boys
- "Little Deuce Coupe" by The Beach Boys
- "I'm in Love with My Car" by Queen
- "Chevy Van" by Sammy Johns
- "Mustang Sally" by Wilson Pickett (mentions a Chevrolet)
- "American Pie" by Don McLean
- "Radar Love" by Golden Earring (mentions a Chevrolet)
- "Pink Cadillac" by Bruce Springsteen (mentions a Chevrolet)
- "Red Barchetta" by Rush (mentions a Chevrolet)

Chevrolet has also contributed to many box office hits at movie theaters. Those include:

- *Transformers* series - Chevrolet Camaro as Bumblebee
- *The Fast and the Furious* series - Chevrolet Corvette
- *American Graffiti* - 1956 Chevrolet Bel Air
- *Better Off Dead* - 1967 Chevrolet Camaro
- *Two-Lane Blacktop* - 1955 Chevrolet
- *Hollywood Knights* - 1957 Chevrolet
- *Corvette Summer* - 1973 Chevrolet Corvette
- *The Runaways* - 1967 Chevrolet Camaro
- *American Hot Wax* - 1957 Chevrolet Bel Air

- *Two for the Money* - 1969 Chevrolet Camaro
- *Christine* - 1958 Chevrolet Bel Air
- *Gone in 60 Seconds* - 1967 Chevrolet Corvette Stingray
- *Baby Driver* - 2004 Chevrolet Impala
- *Drive Angry* - 1969 Chevrolet Chevelle SS

The Chevrolet brand is usually among the top five brands in U.S. car sales markets, month in and month out, year in and year out.

16

BMW

German model of innovation for aircraft engines, cars and motorcycles

Gustaf Otto

"BMW is the epitome of luxury and performance."
—Unknown

BMW, which stands for Bayerische Motoren Werke, Bavarian Motor Works in English, is a German multinational corporation that produces luxury vehicles and motorcycles. BMW was founded as a manufacturer of aircraft engines, but after World War I, it was forced to shift its focus to producing motorcycles and engines for cars.

Three men, Franz Joseph Popp, Karl Rapp and Camilo Castiglioni, founded BMW. Popp was a former general director of the Austro-Daimler company. Rapp was an engineer who had designed engines for aircraft and Castiglioni, an investor. The three men first formed a company, in Munich, Germany, called Rapp Motoren-Werke GmbH in 1913 and began producing aircraft engines.

In 1917, the company changed its name to Bayerische Motoren Werke, or BMW for short. After the end of World War I, BMW was banned from producing aircraft engines by the treaty of Versailles, so it turned its attention to producing motorcycles and engines for cars.

In the 1920s and 1930s, BMW became known for producing high-performance motorcycles and luxury automobiles. Its early successes included the BMW 315, which was the company's first car, and the BMW I-3, which was its first motorcycle. BMW cars were known for their innovative design, precision engineering and exceptional performance.

In the early 1930s, BMW, a company primarily known for its motorcycles and automobiles, recognized the potential of aircraft engines and began developing them. The BMW 801, introduced in the mid-1930s, was designed as a large displacement, air-cooled, 14-cylinder radial engine. It featured a two-stage supercharger and produced around 1,600 to 2,000 horsepower, depending on the variant.

During World War II, BMW produced engines for the German military. After the war, the company focused on motorcycles and cars. In the post-war era. BMW faced financial difficulties, but it still managed to recover and expand its product line.

The BMW 801 was chosen to power the Focke-Wulf Fw 190, a formidable German fighter aircraft designed to outperform its contemporaries. The Fw 190 combined excellent maneuverability, robust construction, and firepower, making it a formidable opponent for Allied aircraft.

The BMW 801 engine played a crucial role in the Fw 190's success. Its power allowed the Fw 190 to achieve impressive speed and climb rates, making it highly effective in combat situations. The engine's reliability and performance at high altitudes also made the Fw 190 a capable interceptor and fighter-bomber. The story of the BMW 801 is intertwined with the development of one of Germany's most iconic fighter planes, the Focke-Wulf Fw 190.

The Fw 190s were such an issue for Allied air superiority, a commando raid was devised to capture one intact at a French airfield and fly it home for evaluation. Luckily this proved to be unnecessary when in 1942 a confused German pilot accidently landed his Fw 190 completely intact at RAF Penbury, an airfield in Wales. The pilot, Armin Faber, incorrectly thought he had landed on a German base in France, even wiggling his wings in celebration as he flew over those on the ground looking up in disbelief. The British inspected the aircraft to the last nut and bolt in order to develop an alternative.

The BMW 801 engine, however, was not without its challenges. It had a complex design and required skilled maintenance. Additionally, the production of the engine faced difficulties due to resource shortages and Allied bombing raids on German factories. Despite these challenges, the BMW 801 engine and the Fw 190 proved to be a formidable combination on the battlefield.

The BMW 801-powered Fw 190s served on various fronts during World War II, engaging in dogfights, ground attacks, and bomber interception missions. The aircraft's superior performance and the engine's reliability contributed to its fearsome reputation.

The Focke-Wulf Fw-190 was widely believed to be the best fighter aircraft of World War II. As the war went on the FW-190 was manufactured in no fewer than 40 different models. The appearance of the new aircraft over France in 1941 was a rude surprise to the Allied air forces.

By the war's end, over 20,000 BMW 801 engines had been produced. The engine's legacy extended beyond its wartime service, as it played a role in post-war aviation development. Some variants of the BMW 801 engine were used in other aircraft, including the Focke-Wulf Ta 152 and the Blohm & Voss BV 155.

The BMW 801 engine stands as a testament to BMW's engineering prowess and its successful foray into the field of aircraft engines. Its association with the Focke-Wulf Fw 190, one of the most iconic German fighter planes of World War II, ensures its place in aviation history.

In the 1960s, BMW introduced its new class of compact cars which helped establish it as a major player in the luxury automobile market. In the 1970s, BMW introduced its first turbo-charged engine and began producing the BMW 3 series, which became one of the company's most successful models.

Today, BMW is one of the world's leading manufacturers of luxury vehicles and motorcycles. The company's product line includes a wide range of cars, SUVs and motorcycles, and it is best known for its commitment to innovation, quality and performance. BMW is headquartered in Munich, Germany, and is led by a team of executives who are committed to keeping the company at the forefront of the automotive industry.

17

Aston Martin

The British are coming

Lionel Martin

*The only way to really understand the beauty of
an Aston Martin is to drive one."
—Jeremy Clarkson, Top Gear*

Aston Martin is a British luxury car company that was founded in 1913 by Lionel Martin and Robert Bamford in London, England. The company was originally called Bamford & Martin Ltd. and produced high-performance racing cars. The name Aston Martin was later adopted after Lionel Martin successfully raced a car at the Aston Hill Climb in Buckinghamshire.

During the 1920s and 1930s, Aston Martin continued to produce racing cars and also began producing luxury touring cars. However, the company struggled financially and went through several ownership changes. In 1947, the company was acquired by David Brown Limited, which also owned the Lagonda car company. Under Brown's ownership, Aston Martin produced some of its most iconic models, including the DB5, which was famously driven by James Bond in the 1964 film *Goldfinger*.

In 1972, Aston Martin was sold to a consortium of investors, but the company continued to struggle financially and went through several more ownership changes over the next few decades. In 1987, Ford Motor Company acquired a 75% stake in Aston Martin, and in 1994, Ford became the sole owner of the company. Under Ford's ownership, Aston Martin saw a resurgence in popularity and produced several successful models, including the DB7 and the Vanquish.

In 2007, Ford sold Aston Martin to a consortium of investors led by Pro-drive founder David Richards. Since then, Aston Martin has continued to produce high-performance luxury sports cars, including the DB11 and the Vantage. In 2018, Aston Martin went public with an initial public offering on the London Stock Exchange.

Television shows and movies that have featured Aston Martin cars include:

- *James Bond* films (various models throughout the franchise)
- *The Avengers* (1961-1969) - featuring the iconic DB5
- *The Persuaders!* (1971-1972) - starring Roger Moore and Tony Curtis, featuring the DBS
- *The Saint* (1962-1969) - starring Roger Moore, featuring the DB4 and DBS
- *The Cannonball Run* (1981) - featuring the V8 Vantage
- *The Living Daylights* (1987) - featuring the V8 Vantage Volante
- *Batman Begins* (2005) - featuring the DB5

Famous people who have owned Aston Martin cars over the years include:

- James Bond (fictional character)
- David Brown (owner of Aston Martin from 1947-1972)
- Peter Sellers (actor)
- Paul McCartney (musician)
- Rowan Atkinson (actor)
- Daniel Craig (actor, also known for playing James Bond)
- Tom Brady (American football player)
- David Beckham (former footballer)
- King Charles III (British royal)

Aston Martin has had a tumultuous history, with many ownership changes and financial struggles. However, the company has remained a symbol of British luxury and craftsmanship, producing some of the remarkably iconic and desirable cars in the world.

18

Maserati

Alfieri Ettore and Ernesto create namesake

Alfieri Ettore

"Maserati is about style, elegance and performance. It's not about being the fastest, it's about being the most beautiful."
—*Luca di Montezemola*

The Maserati car company has a rich history dating back to 1914 when it was founded in Bologna, Italy by Alfieri Maserati and his brothers Ettore and Ernesto. The company started as a small garage that specialized in repairing and modifying cars. However, the Maserati brothers had a passion for racing and soon began building their own race cars.

In 1926, the Maserati brothers moved their company to Modena, Italy, and began producing their own cars under the name Maserati. Their first car, the Tipo 26, was a success on the racing circuit and helped establish the Maserati brand as a serious contender in the racing world.

Over the years, the Maserati company changed ownership several times. In 1937, the company was sold to the Orsi family, known for their successful business ventures in the steel industry. Under the Orsi family's ownership, Maserati continued to produce successful race cars and expanded into producing luxury cars for consumers.

In 1968, the French car company Citroen acquired a majority stake in Maserati. This partnership led to the development of the Maserati Bora, which was a successful luxury sports car that helped boost the company's sales.

The partnership with Citroen was short-lived, and in 1975, Maserati was sold to Alejandro de Tomaso, an Argentinian entrepreneur who also owned the De Tomaso car company. Under de Tomaso's ownership, Maserati continued to produce high-performance sports cars and expanded into producing luxury sedans.

In the 1990s, Maserati faced financial difficulties and was acquired by the Fiat Group. Fiat invested heavily in the company and helped modernize its production facilities. In 1997, Maserati introduced the 3200 GT, which was a successful luxury sports car that helped revive the brand.

Movies and television shows that have featured Maserati cars include:

- *The Italian Job* (1969)
- *Miami Vice* (TV series, 1984-1989)
- *The Sopranos* (TV series, 1999-2007)
- *Entourage* (TV series, 2004-2011)
- *The Dark Knight* (2008)
- *Fast and Furious 6* (2013)
- *John Wick: Chapter 2* (2017)
- *Bad Boys for Life* (2020)

Some famous people who own Maserati cars include:

- Cristiano Ronaldo
- David Beckham
- Jay-Z
- Eddie Murphy
- Gerard Butler
- Shaquille O'Neal
- Floyd Mayweather, Jr.
- Lewis Hamilton
- Jeremy Clarkson (host of *Top Gear* and *The Grand Tour*)
- Jodie Kidd (model and TV personality)

Today, Maserati is owned by Stellantis. The company continues to produce high-performance sports cars and luxury sedans that are known for their distinctive Italian design and powerful engines.

The Maserati car company has a long and storied history that has been marked by several ownership changes. The company has remained true to its roots as a producer of high-performance sports cars and luxury vehicles. Though with its distinctive Italian design and powerful engines, Maserati is a brand that is synonymous with luxury and performance.

19

Dodge

Unlikely duo, Dodge brothers, start amazingly successful automaker

Dodge brothers Horace and John

"The Dodge brothers were the first to put a steel top on their cars and the first to build cars with a sliding gear transmission."
—Henry Ford

The Dodge brothers, John Francis and Horace Elgin, were born in Niles, Michigan, in 1864 and 1868, respectively. They were the sons of a machinist and grew up working in their father's machine shop. In 1886, they moved to Detroit and began working in various machine shops and factories, honing their skills as machinists and the mechanics.

In 1900, the Dodge brothers founded their own machine shop, which specialized in producing parts and assemblies for the fledging auto industry. Their reputation for quality workmanship and reliability quickly grew and they began supplying parts and services to several major automakers, including Oldsmobile, Ford and Packard.

The Dodge brothers relationship with Henry Ford was particularly important as they became one of Ford's largest suppliers of engines, transmissions and other component parts. This partnership proved to be incredibly lucrative for the Dodge brothers as they built a thriving business based on their expertise in manufacturing and production.

However, in 1913 the relationship between Henry Ford and the Dodge brothers began to sour as Ford was planning to build a new state-of-the-art manufacturing facility in Highland Park, Michigan, which would allow Ford to produce cars at a much larger scale than ever before. That caused the Dodge brothers to be concerned that this move would give Ford too much control over their business, and they demanded that he buy them out of their contract. Henry Ford refused.

The Dodge brothers decided to strike out on their own. In 1914, they founded the Dodge Brothers Motor Company, with the goal of producing their own line of cars and trucks. They quickly gained the reputation for building durable, high-quality vehicles and their company grew rapidly.

In the mid-1920s, the Dodge Brothers Motor Company was one of the largest and most successful auto manufacturers in the world. In fact, in 1925 the company was sold to the investment firm Dillon, Reed and Company for the staggering sum of more than $146 million, making it the largest corporate financial transaction in US history at the time. Dodge was sold again just three years later, 1928, to Chrysler Corporation.

Unfortunately, the Dodge brothers did not live to see the full extent of their success. In 1920, John Dodge died of pneumonia while on a trip to Europe. Horace Dodge took over as president of the company, but his health began to decline soon thereafter. In 1926, Horace died of cirrhosis of the liver at the age of 52.

Many people have speculated about the cause of the Dodge brothers' premature deaths. Some have suggested they were heavy drinkers and that this may have contributed to their health problems. Others have pointed to the influenza pandemic of 1918 to 1919 which killed millions of people around the world, including many in the United States.

While it is impossible to say for certain what caused the Dodge brothers' early deaths, their contributions to the auto industry were very profound. Their expertise in manufacturing and production helped to revolutionize the way that cars are made, and their commitment to quality and reliability set a standard that many other automakers would strive to emulate.

Today, the Dodge automotive brand is still one of the most recognizable names in the auto industry. While the company has gone through many changes over the years, it continues to produce some of the most powerful and iconic vehicles on the road, including the Dodge Challenger and Dodge Charger, which are still popular muscle cars.

20

Lincoln

Lincoln Motor Company – created to lead in the luxury vehicle class

Henry LeLand

"The company's future is as bright as the promises of God."
—*Henry Leland, founder of Lincoln and Cadillac*

Henry M. Leland was a prominent figure in the early 20th century automobile industry. He was known for his engineering prowess and his dedication to producing high-quality vehicles. In 1917, Leland and his son, Wilfred, embarked on a new venture by starting the Lincoln Motor Company, also known simply as Lincoln.

This chapter explores the origins of the company, its important role in producing jet engines during World War I, and its eventual acquisition by Henry Ford and Ford Motor Company after facing bankruptcy in 1922.

Henry Leland had already made a name for himself in the automotive industry before founding Lincoln Motor Company. He was one of the co-founders of Cadillac Motor Company and played a significant role in establishing it as a luxury car brand. However, due to a disagreement with Billy Durant, president of General Motors, Leland left Cadillac and decided to start his own company. Thus, the Lincoln Motor Company was born.

The timing of the company's establishment was significant, as it coincided with the United States' entry into World War I. Recognizing the need for advanced technology in the military effort, Lincoln Motor Company then shifted its focus from producing automobiles to manufacturing aircraft engines for the war. This decision proved to be crucial, as the company successfully developed Liberty engines, which were used in various aircraft during the war.

The Liberty engine was renowned for its reliability and power, contributing significantly to the Allied forces' capabilities. Despite the success during the war, Lincoln Motor Company faced financial challenges in the post-war period. The demand for aircraft engines decreased significantly as the war came to an end, and the company struggled to transition back to producing automobiles. These difficulties led to bankruptcy in 1922, leaving the future of the company uncertain.

An unexpected turn of events occurred when Henry Ford, the founder of Ford Motor Company, decided to purchase Lincoln Motor Company out of bankruptcy. Ford recognized the potential and value of the brand, as well as the engineering expertise of Henry M. Leland. Henry Leland and Henry Ford didn't always get along well, though they each respected the other's engineering and personal capabilities.

With the acquisition, Ford aimed to expand its luxury car offerings and compete with other high-end automotive manufacturers.

Under Ford's ownership, Lincoln Motor Company experienced a revival. The company introduced new models that combined Ford's manufacturing efficiency with Leland's commitment to quality and luxury. The Lincoln brand became synonymous with elegance, attracting a wealthy clientele and establishing itself as a formidable competitor in the luxury car market.

Major innovations led by Lincoln over the years include:

- Introduction of the Lincoln Continental: The Lincoln Continental, introduced in 1939, was a flagship luxury car that became an iconic symbol of American automotive design. It featured a sleek and elegant design, luxury amenities and advanced engineering for its time.

- Development of the first V12 engine: In 1932, Lincoln introduced one of the earliest V12 engines in a production car. This engine offered smooth and powerful performance, setting a new standard for luxury automobiles.

- Introduction of the powered convertible rooftop: In the 1940s, Lincoln was one of the first automakers to introduce an electrically powered convertible roof. This innovation allowed for easy and convenient operation of the convertible top.

- Introduction of the personal luxury vehicle: In the 1950s, Lincoln introduced the concept of a personal luxury car with the launch of the Lincoln Mark II. This car offered a combination of luxury, style, and performance, catering to affluent buyers who sought a more individualistic driving experience.

- Introduction of the personal safety system: Lincoln pioneered the Personal Safety System in the 1990s, which integrated multiple safety features like dual-stage airbags, seat belt pretensioners and crash severity sensors. This system aimed to provide enhanced protection for occupants during collisions.

- Development of the SYNC infotainment system: Lincoln, along with Ford, developed the SYNC infotainment system in collaboration with Microsoft. Introduced in 2007, SYNC allowed for voice-activated control of various functions like music, phone and navigation, revolutionizing in-car connectivity.

- Introduction of the Lincoln Black Label: In 2014, Lincoln launched the Black Label program, offering personalized luxury experiences for customers. This program included exclusive design themes, premium services, and enhanced ownership benefits, elevating the brand's luxury image.

- Embracing hybrid and electric technologies: Lincoln has been actively embracing hybrid and electric technologies in recent years. The brand introduced the Lincoln Aviator Grand Touring, a plug-in hybrid SUV, and has plans to release more electric and hybrid models in the future, reflecting a commitment to sustainability and innovation.

The story of Lincoln Motor Company is one of resilience and reinvention. Founded by Henry M. Leland and his son Wilfred in 1917, the company played a crucial role in producing jet engines for the military effort during World War I. Despite facing bankruptcy in 1922, Lincoln Motor Company was saved by Henry Ford's acquisition, leading to its revival as a luxury car brand. The merger between Lincoln and Ford Motor Company proved to be a successful relationship, cementing Lincoln's position as a prominent brand and player in the automotive industry.

21

Mitsubishi

Mitsubishi makes its unique mark on automotive sector

Mitsubishi

"Mitsubishi has always been a company that leads with innovation. We are proud to be at the forefront of technology and design."
—Tetsuro Aikawa

Mitsubishi has a rich and intriguing history as an automaker, with its origins tracing back to its parent company, Mitsubishi Group. In this chapter, we delve into the history of Mitsubishi as an automaker, its entry into the United States markets and its sales trend over the years.

Mitsubishi Group, also known as the Mitsubishi Conglomerate, is a Japanese multinational conglomerate that has its roots in the shipping business. Founded in 1870 by Yataro Iwasaki, the Mitsubishi Group expanded its operations into various industries, including mining, banking and eventually into manufacturing cars. In 1917, Mitsubishi Motors Corporation was established as a division of the Mitsubishi Shipbuilding Company, with the aim of producing cars.

Before getting cars rolling, Mitsubishi was heavily involved in engineering aviation and aeronautics. The Mitsubishi A6M Zero, often referred to simply as the "Zero," was a legendary Japanese fighter plane that played a significant role in World War II. Its creation is a fascinating story that showcases the Japanese pursuit of technological superiority in aviation during the early 20th century.

In the late 1930s, as tensions rose between Japan and other world powers, the Japanese Imperial Navy recognized the need for a new carrier-based fighter aircraft that could outperform existing models. The Navy issued a requirement for a highly maneuverable, long-range, and lightweight fighter plane that could maintain air superiority over the vast expanses of the Pacific Ocean.

The development of the Zero was led by Jiro Horikoshi, an aeronautical engineer at Mitsubishi Heavy Industries, Ltd. He was tasked with designing an aircraft that met the Navy's demanding specifications. Horikoshi and his team worked tirelessly to create a plane that was not only fast and agile but also had an extended operational range.

One of the key innovations in the Zero's design was its use of lightweight materials. The aircraft's frame was constructed from aluminum and other lightweight alloys, allowing it to have a remarkable power-to-weight ratio. Additionally, the Zero incorporated an advanced aerodynamic design with a low drag coefficient, giving it exceptional speed and maneuverability.

To ensure the Zero's long-range capabilities, the design team focused on optimizing its fuel efficiency. They implemented features like a streamlined fuselage and efficient wing design, enabling the aircraft to cover longer distances without refueling.

By 1939, the first prototype of the Mitsubishi A6M Zero took to the skies for its maiden flight. The aircraft demonstrated exceptional performance, exceeding the expectations of the Japanese Navy. The Zero's top speed, climbing ability and maneuverability were superior to most of the fighter planes in service at the time.

The Zero quickly gained a reputation for its dominance in aerial combat, especially during the early stages of World War II. It was known for its incredible agility and range, allowing Japanese pilots to engage enemies at a distance and escape unfavorable situations. The Zero's performance in the Pacific Theater, particularly during the attack on Pearl Harbor and subsequent battles, solidified its status as one of the most iconic fighter planes of its era.

The Mitsubishi A6M Zero played a significant role in several famous World War II battles in the Pacific Theater. Some of the most notable battles where the Zero was involved include the attack on Pearl Harbor (December 7, 1941). The Zero was one of the primary aircraft used by the Japanese in their surprise attack on the U.S. Pacific Fleet stationed at Pearl Harbor, Hawaii. The Zero's agility and speed helped it achieve air superiority during the attack, allowing Japanese forces to devastate the American naval and air assets.

Battle of Coral Sea (May 4-8, 1942): The Zero played a key role in this naval battle between Japanese and Allied forces. It engaged in dogfights with Allied aircraft and was instrumental in sinking the aircraft carrier USS *Lexington,* while another aircraft carrier, the USS *Yorktown,* was damaged.

The Mitsubishi A6M Zero remains a symbol of Japanese aviation engineering and the early successes of the Japanese military in World War II. The aircraft's creation is a testament to the ingenuity and dedication of Jiro Horikoshi and his team at Mitsubishi, who designed a fighter plane that left an indelible mark on history.

Mitsubishi's first venture into the automotive industry was the production of the Model A, a seven-seat sedan. The Model A was introduced in 1917, making Mitsubishi one of the oldest carmakers in Japan. Due to the economic downturn during World War I and subsequent financial difficulties, Mitsubishi temporarily halted its car production.

It wasn't until the 1960s that Mitsubishi made a significant impact as an automaker. The company focused on developing compact cars that were fuel-efficient and affordable, catering to the needs of the post-war economy. Mitsubishi's first breakthrough came with the launch of the Mitsubishi 500, a compact car that quickly gained popularity in Japan. This success laid the foundation for Mitsubishi's expansion into international markets, including the United States.

As for its entry into the US market in the early 1970s, Mitsubishi and Chrysler Corporation embarked on a joint venture that would change the automotive landscape at the time. This collaboration resulted in the production of iconic vehicles such as the Dodge Colt and Plymouth Champ, which later evolved into the Dodge D50, Plymouth Arrow mini pickups, and ultimately, the Mitsubishi Mighty Max.

The joint venture was a strategic move by both companies to leverage each other's strengths and gain a foothold in the competitive American automotive market. At the time, Mitsubishi was a well-established Japanese automaker known for its engineering prowess and fuel-efficient vehicles. On the other hand, Chrysler had a strong presence in the American market and a wide distribution network.

The first fruits of this partnership were the Dodge Colt and Plymouth Champ, which were introduced in 1971. These compact cars were based on Mitsubishi's Galant model and quickly gained popularity for their affordability, reliability and fuel efficiency. The Colt and Champ were available in various body styles, including sedans, coupes and wagons, catering to a wide range of customer preferences.

As the collaboration continued, both companies recognized the growing demand for compact pickup trucks in the American market. In response, they introduced the Dodge D50 and Plymouth Arrow mini pickups in 1979. These vehicles were based on Mitsubishi's existing truck platforms but were modified to suit American tastes and preferences. The D50 and Arrow offered a combination of utility, versatility, and fuel efficiency, making them attractive options for both personal and commercial use.

The joint venture reached its pinnacle with the introduction of the Mitsubishi Mighty Max in 1982. This pickup truck was essentially a rebadged version of the Dodge D50 and Plymouth Arrow, but with Mitsubishi's branding. The Mighty Max carried forward the same attributes that made its predecessors successful, such as reliability, affordability and fuel efficiency. It gained a loyal following among truck enthusiasts and those seeking a dependable workhorse.

The Mitsubishi-Chrysler joint venture was a testament to the benefits of collaboration in the automotive industry. It allowed both companies to capitalize on their respective strengths and create a series of vehicles that resonated with American consumers. The partnership not only helped Mitsubishi establish a strong presence in the American market but also provided Chrysler with a lineup of fuel-efficient and reliable vehicles during a period of economic uncertainty.

However, like many joint ventures, the partnership between Mitsubishi and Chrysler Corporation eventually came to an end. As the automotive landscape evolved, both companies pursued independent paths, and their collaboration faded away. Nevertheless, the legacy of their joint venture lives on in the form of the Dodge Colt, Plymouth Champ, Dodge D50, Plymouth Arrow and Mitsubishi Mighty Max, which continue to hold their special place in automotive history.

Mitsubishi entered the US car markets more directly in 1982 with the introduction of the Mitsubishi Tredia and Cordia models. These compact cars were well-received, offering a combination of affordability, fuel efficiency and reliability. Though it was the launch of the Mitsubishi Eclipse in the late 1980s that truly put the company on the map in the US. The Eclipse, a sporty coupe, appealed to a younger demographic and gained its own near cult following.

Throughout the 1990s, Mitsubishi continued to expand its product line in the US, introducing models such as the Galant, Montero and Diamante. The company also made a name for itself in the performance car segment with the introduction of the Mitsubishi Lancer Evolution, a high-performance sedan that became synonymous with rally racing. Mitsubishi's sales in the United States peaked in the early 2000s, with the popularity of its SUVs and performance cars driving its success.

Mitsubishi faced challenges in the late 2000s and early 2010s. The company was hit hard by the global financial crisis and struggled to keep up with changing consumer preferences. Mitsubishi's sales in the US declined, and the company had to reevaluate its strategy. In recent years, Mitsubishi has shifted its focus towards electric vehicles and SUVs, aiming to capitalize on the growing demand for environmentally friendly and spacious vehicles.

Despite the challenges, Mitsubishi has shown resilience and determination to regain its position in the United States markets. The company has introduced the Mitsubishi Outlander PHEV, a plug-in hybrid SUV, which has gained popularity among eco-conscious consumers. Additionally, Mitsubishi has partnered with Japanese automaker Nissan and European car company, Renault, another Japanese automaker, to share resources and technologies, which has helped improve its product lineup and expand its presence in the United States.

On a personal experience note, our daughter's first car was a new Mitsubishi Eclipse we bought in time for her to drive off to college. Our son's first car was a new Mitsubishi Mirage, which was also acquired before his time heading off to his first year of college. Both cars were solid and reliable vehicles. They performed well and were popular cars for the kids during their early driving years.

In recent years, there has been speculation as to whether Mitsubishi is long for the US market. The number of franchised dealerships carrying the Mitsubishi brand has been drastically reduced, along with the volume of Mitsubishi brand car sales in the US in general.

Mitsubishi Motors' impact on pop culture in the USA has been relatively limited compared to other automakers. While Mitsubishi has had some notable moments in the American market, it hasn't necessarily led any significant pop culture developments. However, here are a few notable instances where Mitsubishi vehicles have appeared in pop culture:

- The Fast and the Furious franchise: Mitsubishi vehicles have featured prominently in the popular movie series, *The Fast and the Furious*. The iconic 1995 Mitsubishi Eclipse GS-T was driven by the character Brian O'Conner, played by Paul Walker, in the first film. This helped to popularize the Eclipse among car enthusiasts.

- Initial D anime and manga: Mitsubishi Lancer Evolution models, particularly the Evolution III and Evolution VIII, are featured prominently in the popular Japanese anime and manga series called Initial D. The series helped to introduce the Lancer Evolution to a wider audience and gained a cult following among car enthusiasts.

- Rally racing success: Mitsubishi has had success in rally racing, particularly with its Mitsubishi Lancer Evolution models. While not directly pop culture developments, these victories and the Lancer Evolution's reputation as a high-performance rally car have gained attention from car enthusiasts and helped to establish Mitsubishi's image in the USA.

It's worth noting that Mitsubishi's influence on pop culture in the US has been relatively limited compared to other automakers like Ford, Chevrolet, or even its Japanese counterparts such as Toyota or Honda.

Mitsubishi's history as an automaker is intertwined with its parent company, Mitsubishi Group. From its humble beginnings as a division of the Mitsubishi Shipbuilding Company, Mitsubishi Motors Corporation has grown into a global player in the automotive industry. Its entry into the United States markets in the 1980s marked a significant milestone, and the company experienced both successes and challenges over the years. Despite the ups and downs, Mitsubishi has continued to innovate and adapt to changing market conditions, ensuring its place as a prominent automaker in the United States.

22

Bentley

British bring luxury front and center with reliable and fast Bentley

W.O. Bentley

"I have always wanted to build a good motor car, a fast car, the best in its class."
—W.O. Bentley

"The Bentley is a fast car. It's just a fast car. It's a car that's like a really fast car."
—Jeremy Clarkson, Top Gear, 2002 - 2015

Bentley Motors Ltd., founded in 1919 by Walter Owen Bentley, also known as W.O. Bentley, started in Cricklewood, North London, with the aim of producing luxury cars that were reliable and fast. The first Bentley car, produced in 1921, was known as the Bentley 3 Litre.

Bentley cars are famous for their luxury, performance and elegance. They are often associated with high-end sports cars and have been used in many prestigious racing events, including the 24 Hours of Le Mans. Bentley cars are also known for their distinctive design, which includes a prominent grille and sleek lines.

In the early years, Bentley cars were primarily used for racing, and the company gained a reputation for producing high-performance vehicles. However, during the 1930s, Bentley shifted its focus to producing luxury cars for wealthy customers. The company continued to produce luxury cars throughout the 20th century, and today, Bentley is known as one of the most prestigious car brands in the world.

Famous movies and television shows that have featured Bentley cars include:

- *James Bond* films: The Bentley Mark IV and the Bentley Mark VI have both been featured in James Bond films.

- *The Avengers:* The Bentley S3 Continental was driven by Emma Peel in the popular British TV series.

- *Downton Abbey:* The show prominently featured a 1920s Bentley 4.5 Litre.

- *Batman Begins:* Bruce Wayne (played by Christian Bale) drives a Bentley Continental GT in the film.

Some famous people who have owned Bentley cars over the years:

- Queen Elizabeth II
- David Beckham
- Jay Z
- Kim Kardashian
- Simon Cowell
- Hugh Grant

- Tom Cruise
- Paris Hilton
- Donald J. Trump
- LeBron James
- Rihanna
- Elton John
- Cristiano Ronaldo
- Justin Bieber
- John Elway
- Floyd Mayweather, Jr.

Ownership of the Bentley brand has changed several times over the years. Here are some of the major changes:

- 1919-1931: Bentley Motors Limited was founded by W.O. Bentley. It was later acquired by Rolls-Royce in 1931.

- 1998: Volkswagen Group acquired Bentley Motors Limited from Rolls-Royce.

- 2002: BMW acquired the rights to the Rolls-Royce brand, but Volkswagen Group retained Bentley.

- 2020: Volkswagen Group announced that Bentley would become a subsidiary of Audi AG, another subsidiary of Volkswagen Group.

Throughout its history, Bentley has faced many challenges, including financial difficulties and changes in ownership. The company has managed to overcome these challenges and maintain its reputation as a leading luxury car manufacturer. Today, Bentley continues to produce high-end cars that are sought after by collectors and enthusiasts around the world.

23

Jaguar

The British performance and luxury car brand goes global

William Lyons

"The Jaguar is simply a beautiful machine, and I couldn't resist the opportunity to own one."
—Sir William Lyons, Co-founder of Jaguar

Jaguar is a British automotive brand that, over the years, has become synonymous with luxury and high performance. The company was founded in 1922 as the Swallow Sidecar Company by William Lyons and William Walmsley. Initially, the company produced sidecars for motorcycles, but it soon expanded to manufacturing automobiles. Over the decades, Jaguar has become one of the most popular and recognizable luxury and performance automotive brands in the world.

The early years of Jaguar were marked by innovation and a commitment to quality. In 1931, the company introduced the SS1, its first car to sport a six-cylinder engine. This was a significant milestone for the company, as it marked the beginning of Jaguar's reputation for producing high-performance vehicles.

During World War II, Jaguar shifted its focus to producing aircraft components and engines. This helped the company gain valuable experience in engine design and manufacturing. After the war, Jaguar returned to producing automobiles, and in 1948 it introduced its most famous model, the XK120. This car was a sensational hit and it quickly established Jaguar as a leader in the automotive industry.

Throughout the 1950s and 1960s, Jaguar continued to produce high-performance vehicles that were both stylish and reliable. In 1961, the company introduced the E-Type, which was widely regarded as one of the most beautiful cars ever made. The E-Type was a huge success, and it helped solidify Jaguar's reputation as a manufacturer of high-performance sports cars.

During the 1970s, Jaguar faced a number of challenges. The company was struggling financially, and it was forced to merge with the British Motor Corporation (BMC) to stay afloat. This merger resulted in the formation of British Leyland, which was a con-glomerate of several British automakers. Despite these challenges, Jaguar continued to produce high-quality vehicles, including the XJ6, which was introduced in 1968 and remained in production until 1992.

The 1980s proved to be a period of significant growth and expansion for Jaguar. In 1984, the company was privatized and sold to Ford Motor Company. This allowed Jaguar to access Ford's resources and expertise, which helped the company develop new models and expand its global reach.

During the 1990s, Jaguar continued to produce high-quality vehicles, including the XJ220, which was one of the fastest cars of its time. However, the brand faced increasing competition from other luxury automakers, and it struggled to maintain its share in a tough market. In 2008, Ford Motor Company sold Jaguar and its companion brand, Land Rover, to Tata Motors, an India-based automotive company, which has continued to make financial investments in the brand and develop new models.

Jaguar is known for producing a range of luxury vehicles, including sedans, coupes and SUVs. The company's current lineup includes the F-Type, a high-performance sports car, the XE, a compact luxury sedan, and the F-Pace, a luxury SUV. Jaguar has also embraced electric and hybrid technology, and it has developed several electric and hybrid vehicles, including the I-Pace, which is an all-electric SUV. I was fortunate to drive the first Jaguar I-Pace in Colorado between two Jaguar dealerships. At about 60 miles, it was one of the longest drives I had experienced at the time in an all-electric vehicle.

Jaguar's success over the years can be attributed to several factors. First, the company has always been committed to producing high-quality vehicles that are both stylish and reliable. This commitment to quality has helped Jaguar establish a reputation for excellence in the automotive industry.

Second, Jaguar has been able to adapt to changing market conditions and consumer preferences. The company has introduced new models and technologies as needed, and it has been able to stay ahead of the curve in terms of design and innovation.

Finally, Jaguar has been able to leverage its heritage and brand recognition to build a loyal customer base. The company's iconic designs and high-performance vehicles have helped it establish a strong brand identity, which has helped it maintain its market share over the years.

Jaguar is a British automotive brand that has become synonymous with luxury and high performance. Over the years, Jaguar has produced a range of iconic vehicles, including the XK120, the E-Type and the

XJ6. Today, Jaguar is known for producing a range of luxury vehicles, including sedans, coupes and SUVs, and it has embraced electric and hybrid technology. Jaguar's success can be attributed to its commitment to quality, its ability to adapt to changing market conditions and its strong brand identity.

24

Volvo

As a safety and innovative Swedish car maker, Volvo spawns new auto startup Polestar

Gustaf Larson

"Volvo is a Swedish car company known for its commitment to safety and innovation."
—*Unknown*

Volvo, a Swedish automaker, has been producing high-quality vehicles since 1927. The company has a reputation for safety and reliability, and its cars are known for their durability and longevity. Volvo has a rich history, and its success can be traced back to its early years.

The start of Volvo dates to 1924, when a group of Swedish businessmen decided to create a new car company. The group, which included Gustav Larson and Assar Gabrielsson, had a vision of creating a car that was safe, reliable and affordable. They also wanted to build a car that would be able to withstand the harsh Swedish climate and the rough terrain of the country's rural areas.

The first car built by Volvo was the OV 4, introduced in 1927. The car was a success, and it quickly became popular in Sweden. The OV 4 was followed by a series of other cars, including the PV 651 in 1929, a midsized car designed for families. It was one of the first cars to feature a four-speed manual transmission.

Despite the success of its early vehicles, Volvo faced financial difficulties in the 1930s. The company was forced to merge with another Swedish automaker, and it became part of the SKF group. However, Volvo continued to innovate, and it introduced several new technologies during this time. In 1937, Volvo introduced the first car with a laminated safety glass windshield, which was a major breakthrough at that time for automotive safety.

During World War II, Volvo focused on producing military vehicles and equipment. After the war, the company returned to producing cars. In the 1950s, Volvo introduced several new models, including the PV 444 and the PV 544, designed to be affordable and practical. They quickly became popular in Sweden and other parts of Europe.

In the 1960s, Volvo began to focus on safety as a key selling point for its cars. The company introduced a number of new safety features, including three-point seat belts, (invented by Nils Bohlin in 1959, this safety feature has since become standard in cars worldwide), padded dashboards and collapsible steering columns. These features helped to establish Volvo as a prominent leader in automotive safety.

During the 1970s, Volvo continued to innovate, introducing a number of new models, including the 240 series and the 260 series. These cars were designed to be practical and reliable, and they were popular with families and businesses. In 1974, Volvo introduced the first car with a catalytic converter, a major breakthrough in reducing emissions.

During the 1980s, Volvo continued to focus on safety, introducing a number of new safety features, including anti-lock brakes, airbags and traction control. These features helped to continue the tradition of confirming Volvo as an important industry leader in automotive safety and as a pioneer in collision avoidance systems, known for its commitment to protecting its customers.

In the 1990s, Volvo faced financial difficulties, and it was acquired by Ford in 1999. However, during the time it was owned by Ford, Volvo continued to innovate, and introduced a number of new models, including the S80, the V70, and the XC90.

In 2010, Volvo was acquired by Geely, a Chinese automaker. Even through the corporate headquarters and design center remain in Sweden, it is now a China-owned company Since then, Volvo has continued to innovate, and it has introduced a number of new models, including the XC40 and the XC60. These cars are designed to be practical and reliable, and they feature the latest safety and technology features.

Polestar is a Swedish car company that was founded in 1996 as a racing team for Volvo. In 2015, Volvo acquired Polestar and turned it into its performance division. In 2017, Volvo and its parent company, Geely, announced that Polestar would become a separate brand focused on electric performance cars.

Polestar's first car, the Polestar 1, was unveiled in 2017 and is a hybrid grand touring coupe with a range of up to 77 miles on electric power alone. The Polestar 2, a fully electric sedan, was unveiled in 2019 and is the company's first mass-market car. Polestar plans to introduce additional models in the future, including an electric SUV.

Polestar's cars are designed to be environmentally friendly and are manufactured using sustainable materials and production methods. The company also offers a subscription model for its cars, which includes all maintenance and insurance costs, as well as the ability to upgrade to a new model every few years.

Today, Volvo is known for its commitment to safety, reliability, and innovation. The company continues to produce high-quality cars that are popular with families and businesses around the world. With a rich history and a commitment to excellence, Volvo is bound to continue to be a leader in the global automotive industry for years to come.

25

Chrysler

Walter Chrysler creates a leading automaker

Walter Chrysler

*"Chrysler has always been a company that stands
for quality and innovation."*
—*Walter P. Chrysler, founder of Chrysler*

Walter Chrysler is widely regarded as one of the most influential figures in automotive history. Born in Wamego, Kansas, in 1875, Chrysler grew up in a modest family and was forced to leave school at the young age of 17 to support himself. Despite these challenges, Chrysler went on to become one of the most successful entrepreneurs of his time, founding the Chrysler Corporation, and leading it to become one of the Big Three automakers in the United States.

Chrysler began his career as an apprentice in a railroad shop in Ellis, Kansas. He quickly demonstrated a talent for machinery and a strong work ethic, and was soon promoted to foreman. In 1901, Chrysler moved to Pittsburgh, Pennsylvania, and took a job with the American Locomotive Company, where he quickly rose through the ranks to become the company's youngest-ever division superintendent.

Walter Chrysler was recruited, in 1911, by David Buick at the Buick Motor Company, a division of General Motors. He quickly became the key senior executive at Buick and was credited with developing many of the company's most successful models, including the Buick Six. However, Chrysler's ambition exceeded his role at Buick and he resigned from the company in 1919 to pursue his own business ventures.

In 1925, in Highland Park, Michigan, a Detroit suburb, Walter Chrysler founded in his namesake, the Chrysler Corporation from the remains of Maxwell Motor Company and quickly became one of the most successful automakers in the United States. Under Chrysler's leadership, the Chrysler Corporation grew quickly and significantly, and the company introduced a number of ground-breaking automotive innovations. Those include the first practical hydraulic brakes and the first mass-produced cars with four-wheel hydraulic brakes. Chrysler also pioneered the use of wind tunnel testing, which allowed the company to design more aerodynamic vehicles and improve fuel efficiency.

By 1928, the Dodge brand was already well-established and had already developed a loyal following. The acquisition of Dodge Brothers allowed Chrysler to expand the company's offerings and increase market share. The move was a success and Dodge quickly became one of Chrysler Corporation's most successful new car brands.

Building the Zeder

The Zeder model was a remarkable piece of engineering designed by key Chrysler engineers in the early 1930s. It was the first car to be fully developed in the company's newly established engineering department, and it served as a prototype for many of the iconic Chrysler vehicles that would follow. The Zeder was a significant achievement for the Chrysler team and a testament to their commitment to innovation and excellence.

The Zeder was designed by three key engineers at Chrysler: Owen Skelton, Carl Breer and Fred Zeder. These three professional engineers were instrumental in establishing the company's engineering department and were committed to creating a car that was both functional and stylish. It was their belief that the future of the automobile industry rested in aerodynamics and lightweighting of materials and they set out to create a car that embodied these principles.

The Zeder was built on a custom-made chassis and featured a streamlined body that was designed to reduce wind resistance and improve fuel efficiency. The Zeder was also considerably lighter than other vehicles of its time, thanks to its extensive use of aluminum and other lightweight materials. The Zeder was powered by a six-cylinder engine that was capable of producing 70 horsepower, which was quite impressive for a car of its size and weight.

One of the most significant features of the Zeder was its front-wheel drive system. This was a revolutionary innovation at the time which allowed the car to be more maneuverable and easier to handle than other vehicles of its era. The front-wheel drive system also contributed to the car's overall weight reduction, as it eliminated the need for a heavy driveshaft and differential.

The Zeder was first unveiled to the public in 1932 and it proved to be an instant sensation. Its sleek, aerodynamic design and advanced engineering features captured the imagination of car enthusiasts and industry experts alike. The car was praised for its performance, handling and fuel efficiency, and it set a new standard for automotive design and engineering.

The Zeder was an important milestone in the history of Chrysler and the automotive industry. It demonstrated the company's commitment to innovation and excellence, and it set the stage for many of the iconic Chrysler models that would follow. Though it was never mass produced for consumer accessibility, the Zeder was a testament to the talent and dedication of the key Chrysler engineers who designed it. It remains a remarkable automotive achievement.

Beyond the Zeder at Chrysler

Through the 1930s, Chrysler continued to expand the company's offerings, introducing the Plymouth and DeSoto brands to complement the flagship Chrysler brand. During this timeframe, Chrysler also developed a number of new technologies, including the first fully automatic transmission and the first practical air conditioning system for automobiles.

Chrysler adds the Imperial as luxury model

Also in the 1930s, Chrysler began to produce luxury cars under the Chrysler brand. These cars were designed to compete directly with the other notable luxury brands Cadillac and Lincoln, produced by General Motors and Ford, respectively, which were the dominant US luxury car brands at the time. The Chrysler Imperial was introduced in 1931 as a top-of-the-line Chrysler model. It was a large, luxurious car that was designed to appeal to the wealthiest buyers.

The Imperial was redesigned in 1934, and the new model was even more luxurious than the previous one. The car was powered by a high performance eight-cylinder engine and included a number of advanced features, such as hydraulic brakes. The Imperial was also one of the first cars to feature air conditioning as standard equipment, which was a major selling point for buyers in hot climates.

In the 1940s, the Imperial brand was temporarily discontinued due to the World War II at a time almost all new car production in the United States ended. Chrysler shifted its focus to producing military vehicles and other war-related products. The Imperial brand was revived in 1946, and the new model was again designed to compete with Cadillac and Lincoln.

The post-war Imperial was a large, luxurious car that featured a number of advanced features, including power windows, power seats and a push-button radio. The car was powered by a commanding eight-cylinder engine and was available in a number of body styles, including a sedan, a coupe and a convertible.

The new Imperial was designed to compete with Cadillac and Lincoln, which had both undergone major redesigns in the 1950s. The Imperial was marketed as a car that was more modern and stylish than its competitors, and it quickly became popular with buyers who were looking for a luxury car that was both elegant and powerful.

In 1955, the Imperial brand was realigned internally into a separate division of the Chrysler Corporation called the Imperial Division. It was tasked with producing luxury cars that were marketed under the Imperial brand name. The first car to be produced by the new division was the 1955 Imperial, which was designed to compete with the Cadillac Eldorado and the Lincoln Continental Mark II.

The 1955 Imperial was a large, luxurious car that was powered by a high-performance V-8 engine. The car was available in a number of body styles, including a sedan, a coupe and a convertible. The car was also available with a number of advanced features, including power windows, power seats and a push-button-controlled automatic transmission.

The Imperial brand continued to produce luxury cars throughout the 1950s and 1960s. The cars were known for their elegant styling, luxury features and powerful engines. The brand was also known for its innovative technology, including the first electronic fuel injection system, which was introduced in 1958.

In the 1960s, the Imperial brand underwent another major redesign. The newest model was significantly larger than the previous models and featured a number of advanced features, including power windows, power antenna and a six-way power seat.

Major US car manufacturers consolidate

In 1987, Chrysler acquired the struggling automaker, American Motors Corporation (AMC). The acquisition of AMC was particularly significant as it allowed Chrysler to expand its offerings to include the popular Jeep brand of vehicles that AMC had acquired in 1970.

In 1998, Chrysler and its subsidiaries entered into a partnership dubbed a "merger of equals" with German-based Daimler-Benz AG, creating the combined entity DaimlerChrysler AG. To the surprise of many stockholders, Daimler acquired Chrysler in a stock swap before then Chrysler CEO Bob Eaton retired. Under DaimlerChrysler, the company was named DaimlerChrysler Motor Company LLC, with its US operations generally called "DCX."

On May 14, 2007, DaimlerChrysler announced the sale of 80.1% of Chrysler Group to American private equity firm Cerberus Capital Management, LP, there-after known as Chrysler, LLC, although Daimler (renamed as Daimler AG) continued to hold a 19.9% stake.

The economic collapse during the financial crisis of 2008-2009 pushed the company to the brink. On April 30, 2009, the automaker filed for Chapter 11 bankruptcy protection to be able to operate as a going concern, while renegotiating its debt structure and other obligations, which resulted in the corporation defaulting on over $4 billion in secured debts. The US government described the company's action as a "prepackaged surgical bankruptcy".

On June 10, 2009, substantially all of Chrysler's assets were sold to "New Chrysler", organized as Chrysler Group LLC. On January 21, 2014, Fiat bought shares of Chrysler worth $3.65 billion. A few days later, the intended reorganization of Fiat and Chrysler under a new holding company, Fiat Chrysler Automobiles, together with a new FCA logo, were announced.

On January 19, 2021, France-based PSA Group (Peugeot) merged with Fiat Chrysler (FCA). This new company, with 14 brands of cars, immediately became the 4th-highest volume automaker in the world. The company adopted the name Stellantis.

Stellantis has continued to innovate, introduced new technology and has remained one of the most compelling automakers in the world. Key US models produced today by Stellantis under their respective US brands are Chrysler, Dodge, Jeep, Ram and Wagoneer. The most popular models include Pacifica, Challenger, Charger, Wrangler, Cherokee, Grand Cherokee, Grand Wagoneer and Ram Trucks.

Throughout his career, Walter Chrysler was known for his attention to detail, his focus on quality and his commitment to innovation. Chrysler was also a strong advocate for worker's rights and was one of the first automakers to introduce a profit-sharing plan for employees. Chrysler was also a philanthropist and established the Chrysler Foundation, which has supported a wide range of charitable causes.

26

Pontiac

**City in Michigan becomes naming
source for local car brand**

1925 Pontiac

*"Pontiac is a brand that has always stood for passion and
performance and the new GTO is the ultimate
expression of that heritage."*
—Bob Lutz, Vice Chairman of General Motors, 2003

Pontiac was a brand of automobiles manufactured by General Motors (GM), one of the world's largest automakers from, 1926 to 2010. Unlike other brands that made up the early years of GM, including Buick, Oldsmobile, Cadillac and Chevrolet, mostly named for their founders or early inventors/engineers, Pontiac was named for a city near Detroit where early Pontiacs were built.

GM established Pontiac in 1926 as a companion marque to its already operational Oakland brand, and its name was derived from the town of Pontiac, Michigan, where the company was founded. Over the years, Pontiac became known for producing affordable, performance-oriented vehicles that appealed to a wide range of consumers.

The first Pontiac was the Pontiac Series 6-27, which was introduced in the brand's first year, 1926, and was powered by a six-cylinder engine. The car was a success and Pontiac quickly became known for producing affordable, reliable vehicles that were popular with the American public. During the 1930s, Pontiac expanded its line-up to include a range of vehicles, including coupes, sedans and convertibles.

During World War II, Pontiac, like almost all other automakers, shifted its production to support the war effort. After the war, Pontiac resumed production of civilian vehicles, and in the 1950s, the brand began to gain a reputation for producing performance-oriented cars. In 1955, Pontiac introduced its first V-8 engine, which was a huge success and helped to establish the brand as a leader in the performance car market. The 1960s were a golden era for Pontiac, with the introduction of iconic models such as the GTO, Firebird and Grand Prix.

The GTO, often referred to in automotive slang as Goat, was introduced in 1964 as a muscle car that quickly became a cultural phenomenon. The GTO was powered by a V-8 engine and could reach speeds of up to 60 miles per hour in just over six seconds. The GTO was a huge success in performance car circles and it helped to establish Pontiac as a leader in the high-performance car market. The Firebird, which was introduced in 1967, was a pony sports car designed to compete with Ford's Mustang, which launched in 1964. The Firebird was also a success and it helped to establish Pontiac as a leader in the sports car market.

During the 1970s, Pontiac continued to produce performance-oriented cars, but it faced increasing competition from foreign automakers. The oil crisis of the early 1970s also had a significantly negative impact on the automotive industry as consumers began to demand more fuel-efficient cars. In response, Pontiac introduced a range of smaller, more fuel-efficient vehicles, including the Sunbird and the Phoenix.

The 1980s were a challenging decade for Pontiac, as the brand struggled to find its footing in a rapidly changing automotive industry. In 1982, Pontiac introduced the Fiero, a two-passenger sports car that was intended to compete with the Toyota MR2. The Fiero was a unique car, as it was one of the few mid-engine vehicles produced by an American automaker. However, the Fiero was plagued by quality issues and was discontinued in 1988.

During the 1990s, Pontiac shifted its focus away from performance-oriented cars and toward more practical vehicles. In 1992, Pontiac introduced the Grand Am, a mid-size car that was designed to compete with the Honda Accord and the Toyota Camry. The Grand Am was a success, and it helped to establish Pontiac as a leader in the mid-size car market. In 2000, Pontiac introduced the Aztek, a crossover SUV that was designed to appeal to younger consumers. However, the Aztek was a commercial failure, and it is sometimes cited as one of the worst cars ever produced. Ironically, I bought an Aztek for our son who was in college at the time. It was a very reliable and relatively efficient vehicle to operate at the time. Kendall Jackson actually liked the Pontiac Aztek. I did too.

In the 2000s, Pontiac continued to struggle, as the brand faced increasing competition from foreign automakers. In 2004, Pontiac introduced the G6, a mid-size car that was designed to compete with the Honda Accord and the Toyota Camry. The G6 was a decent car, but never captured the imagination of the car-buying public in the way that the GTO and the Firebird had done previously. Ours was a Pontiac G6 convertible in dark blue color.

During the worldwide economic recession of 2008-09, I was part of a lobbying team that advocated for bridge loans for the US domestic automobile manufacturers. CEOs from GM, Chrysler Corporation (then owned by Cerberus Capitol) and Ford Motor Company were all on the team advocating for bridge loans. Only GM and Chrysler ended up seeking the bridge loans. Ford avoided the need for bridge loans by accessing capital before the recession hit. Both GM and Chrysler Corporation eventually went through the bankruptcy process to gain relief from long-term debt. As part of its restructuring plan, GM announced that it would discontinue the Pontiac brand, along with Saturn, Saab and Hummer. Those were dark times for the industry and triggered the demise for Pontiac as a new car brand.

27

Porsche

Ferdinand Porsche creates high-performance and legendary car brand

Ferdinand Porsche

"In the beginning, I looked around and could not find the car I'd been dreaming of: a small, lightweight sports car that uses energy efficiently. So I decided to build it myself."
—Ferdinand Porsche, founder of Porsche

"Porsche is not a car company. It's a lifestyle."
—Magnus Walker, Porsche enthusiast

The Porsche Motor Company is a German automobile manufacturer that specializes in high performance sports cars, both sedans and SUVs. The company was founded by Ferdinand Porsche, a talented engineer and automotive designer who is responsible for some of the most appealing cars of the 20th century.

Ferdinand Porsche was born in 1875 in Maffersdorf, a small town in what is now the Czech Republic. He demonstrated an early aptitude for mechanics and engineering and began working as an apprentice to a local carriage maker at the ripe young age of 14. He later studied engineering in Vienna and worked for several companies in the automotive industry before deciding to strike out on his own.

Ferdinand Porsche was involved in the development of electric cars. He can be credited with creating one of the earliest functional hybrid vehicles.

In the early 20th century, Ferdinand Porsche was working for the Austrian automotive company Lohner. In 1900, he introduced the Lohner-Porsche Mixte Hybrid, which was a groundbreaking vehicle for its time. The Mixte Hybrid featured electric motors in each wheel hub, powered by batteries, along with a gasoline engine that acted as a generator to charge the batteries.

This innovative hybrid system allowed the Lohner-Porsche Mixte Hybrid to operate solely on electricity for short distances, making it one of the earliest electric cars. The gasoline engine provided additional power when needed, effectively extending the vehicle's range.

The Lohner-Porsche Mixte Hybrid garnered attention and accolades, even receiving a Grand Prix award at the 1901 Paris Exposition. It showcased Porsche's forward-thinking approach to alternative propulsion systems and his vision for the future of transportation.

Ferdinand Porsche founded his own engineering consultancy based in Stuttgart, Germany. The company was called Porsche GmbH and its initial focus was on designing and developing high-performance engines for other carmakers. One of Porsche GmbH's biggest clients was the German government, which commissioned Porsche to design a car that would be affordable and practical for German families.

The result of that was the Volkswagen (VW) Beetle (see chapter on Volkswagen), which was introduced in 1938 and quickly became one of the most popular cars in automotive history. The Beetle was designed to be simple, reliable and easy to maintain. The VW Beetle was sold at a price that made it accessible to millions of people around the world.

Despite the success of the Beetle, Ferdinand Porsche was always interested in building high-performance sports cars. During the 1940s, Porsche began working on a prototype for a new car that would be even faster and more powerful than anything available on the market at the time. The result was the Porsche 356 which was introduced in 1948.

The Porsche 356 was a two-seat sports car that featured a lightweight body and carried a rear-mounted engine and a sleek, aerodynamic design. The Porsche 356 quickly gained a reputation as one of the best handling cars on the road, and it won numerous races and rallies around the world.

Over the next few decades, Porsche continued to innovate and push the boundaries of automotive design. On September 12, 1963, the Porsche 911 was unveiled at the Frankfurt International Motor Show and became one of the most iconic sports cars of all time. Full production began a year later in September 1964. The 911 featured a flat six engine, a rear-mounted transmission and a distinctive shape that remained largely unchanged over the next 50-years.

The Porsche 911 was a huge and instant success, and quickly became the flagship model for the company. Over the years, Porsche has introduced numerous variations of the 911, including the turbo, the GT3 and the Targa, each of which was praised for performance and handling.

Ferdinand Porsche once famously said, "I couldn't find the sports car of my dreams, so I built it myself."

In addition to the 911, Porsche introduced a number of other successful models. In the 1970s, the company introduced the Porsche 924, and the 928, both of which were front engine sports cars that were designed to compete with the likes of the BMW 3-series and the Mercedes-Benz SL. During the 1990s Porsche introduced the Boxster, a mid-engine convertible that was aimed at attracting younger, more style-conscious buyers.

Today, Porsche is one of the most successful and respected automakers in the world. The company remains headquartered in Stuttgart, Germany, and it produces a range of high-performance sports cars, SUVs and sedans. Its lineup includes the 911, the Cayman, Panamera, Macan and Taycan, among others. The Taycon is an electric vehicle that was the first EV mass produced by Porsche and has been well received and popular. The massive technological advancements needed for the Porsche Taycan will prove adaptable to other all-electric models that are expected to follow.

28

Nissan

Early Japanese automotive powerhouse

Yoshisuke Aikawa

"Nissan has always been a pioneer in the automotive industry, constantly innovating and leading the way in technology and design."
—*Hirota Saikaw, former CEO of Nissan*

The Nissan Motor Corporation is one of the most well-known and well-respected car manufacturers in the world. The company, based in Japan, has a long and storied history that dates back to its founding in 1933. From its humble beginnings as the Datsun Motor Corporation, Nissan has grown up to be a global powerhouse with a wide range of successful vehicles and a reputation for innovation and quality.

The early history of Nissan is tied closely to the history of the Japanese automotive industry overall. In the 1920s and 1930s, Japan was rapidly industrializing and modernizing. Japan's leaders saw the development of a domestic automobile industry as a key and integral component of this process. In 1931, the Japanese government announced a plan to create a people's car that would be affordable and accessible to ordinary Japanese citizens, regardless of income levels.

In response to this initiative, a group of investors led by Yoshisuke Aikawa founded the Jidosha Seizo Co Ltd in 1933. The company name translates to Automobile Manufacturing Company in English, yet it is better known by its abbreviated name Nissan. Aikawa had previously founded the Tobata Casting Company which supplied parts to the fledgling Japanese auto industry. He saw an opportunity to expand his business by entering the car manufacturing market directly.

Nissan's first car, the Datsun Type 12, was introduced in 1935. It was a small affordable car designed to appeal to the mass market in Japan. It was powered by a 747-cubic centimeter engine that produced just 15 horsepower, but was reliable and fuel efficient, which made it popular with Japanese consumers.

Despite its early success, Nissan faced a number of challenges in the years leading up to World War II. The company struggled to compete with larger and more established Japanese automakers such as Toyota and Mitsubishi. It was further hampered by the economic disruptions caused by the war. Nevertheless, Nissan managed to survive the war and emerge as a major player in the postwar Japanese economy.

In the weeks following the war, Nissan underwent a period of rapid growth and expansion. The company began producing trucks and buses in addition to passenger cars and expanded its operations to other parts of Asia and to Europe. In 1958, Nissan established a joint venture with the British carmaker Austin to produce cars in the UK, and in 1960 established a joint venture with the French automaker, Renault, to produce cars in France.

Not only are Nissans the first mass-produced Japanese vehicles, their unique automotive style made a major impact on the US market when Nissan (then still Datsun) sedans and compact pickups were first imported in 1958. That was also the year Nissan franchised its first US dealerships.

One of Nissan's most successful products during this period was the Datsun 510, introduced in 1967. The Datsun 510 was a compact sedan designed to compete with popular models such as the Volkswagen Beetle and the Toyota Corolla. It was powered by 1.6 Liter engine that produced 96 horsepower and it was known for its responsive handling and excellent fuel economy. The Datsun 510 was a hit with drivers in both the US and Japan and initiated the effort to establish Nissan as a serious contender in the global automotive market.

During the 1970s and 1980s, Nissan continued to expand its product line and its global reach. The company introduced a number of innovative models including the Datsun 240 Z, which proved very popular. The Z-cars were designed to appeal to drivers who wanted a high-performance sports car at an affordable price.

Nissan was one of the first automakers to manufacture and market practical and popular electric vehicles, when it rolled out its now famous Nissan Leaf in 2011. I was present at Empire Lakewood Nissan, March 22, 2011, when the first Nissan Leaf was delivered in Colorado. In 2022, Nissan distributed its first all-electric SUV, the Ariya. Due to the general popularity of SUVs, the Ariya is expected to become extremely popular, driving increased sales.

29

Volkswagen

The people's car for the world

Ferdinand Porsche

"Volkswagen is a miracle."
—Ferdinand Porsche

Volkswagen is one of the most recognizable auto manufacturers in the world, known for producing reliable and efficient vehicles that are accessible to people from virtually all walks of life. The company was founded in Germany in 1937 during a time of significant political and economic upheaval and turmoil in Europe. Despite this challenging environment, a group of key people came together to establish Volkswagen as a corporation and automaker. Their efforts would create a meaningful and lasting impact on the transportation and mobility industry for Germany, Europe and the world.

The founding Volkswagen can be traced back to the early 1930s when the German government began to explore the possibility of building a People's car that would be affordable and accessible for the masses, literally anyone and everyone. At the time, many—even most—new cars were expensive and mostly out of reach of the average consumer of the day. The German government intended to address this by creating a car that would be reliable, inexpensive and super easy to repair when necessary.

In 1934, the German Labor Front appointed Ferdinand Porsche to serve as the designer for the People's car. Porsche, a respected engineer and designer, had worked for several automakers. He was tasked with creating a car that could be produced on a large scale and sold for a price that almost everyone could afford.

Porsche's design for the People's car was completed in 1936 when the first prototype was produced. The following year the Volkswagen factory was converted to produce military vehicles instead of cars. After World War II, the British army took control of the Volkswagen factory handing it back to the German government in 1949. The German government recognized the potential of the Volkswagen as a potentially popular brand and began to develop plans to help turn it into a highly successful auto manufacturer.

Heinrich Nordoff was a successful engineer who worked for Opel before being appointed to the managing director position of Volkswagen in 1948. Under Nordoff's leadership, Volkswagen produced the Beetle sedan or People's car Volkswagen, which was also called internally "Type 1." Apart from the introduction of the Volkswagen Type 2 commercial vehicle (van, pickup and camper), and the VW Karmann Ghia sports car, Nordhoff generally pursued the one-model policy until shortly before his death in 1968.

Nordoff was known for his commitment to quality and innovation, and he oversaw a number of important developments at Volkswagen during his tenure as managing director. Nordoff introduced new manufacturing techniques and streamlined production processes that helped increase efficiency and reduce costs. Nordoff also invested heavily in research and development, led to the creation of new models and features that would set Volkswagen apart from its competitors.

Another key figure in the founding of Volkswagen was Ivan Hurst, a British army officer responsible for overseeing the Volkswagen factory after the war. Hurst recognized the potential of the Volkswagen brand and worked tirelessly to ensure the factory was able to produce high-quality vehicles that could compete with other automakers. Hurst was instrumental in getting the Volkswagen factory back up and running after the war and he played a key role in ensuring that the Beetle became a global phenomenon. He was also responsible for introducing new marketing strategies that helped increase the visibility of the Volkswagen brand and attract new customers.

The two highest-impact cars from Volkswagen were the Beetle (People's car) and the VW Bus. They both ended up in popular movies.

The Volkswagen Beetle became a popular car in the United States during the 1960s, and it was often seen as a symbol of counter-culture and rebellion. In the movies, the Volkswagen Beetle was often portrayed as a lovable and quirky car with a personality of its own.

In the movie *The Love Bug* and its sequels, the Beetle was given a name (Herbie) and was portrayed as a sentient car with a mind of its own. Herbie was a mischievous and fun-loving car that would often get into humorous situations, and the movies were generally lighthearted and comedic.

In the *Transformers* movies, the Volkswagen Beetle was used as a nod to the original Bumblebee character from the *Transformers* franchise. Bumblebee was a Volkswagen Beetle in the original cartoon and comic book series, and the filmmakers decided to pay homage to this by including a yellow Volkswagen Beetle in the movies.

Overall, the Volkswagen Beetle was often used in movies as a symbol of fun, quirkiness and nostalgia. Its unique design and history made it a popular choice for filmmakers looking to create memorable and iconic cars on screen.

- *The Love Bug* – 1968
- *Herbie Rides Again* – 1974
- *Herbie Goes to Monte Carlo* – 1977
- *Herbie Goes Bananas* – 1980
- *Transformers* – 2007
- *Transformers: Revenge of the Fallen* – 2009
- *Transformers: Dark of the Moon* – 2011
- *Transformers: Age of Extinction* – 2014
- *Avengers: Age of the Ultron* – 2015
- *Transformers: The Last Knight* – 2017

The Volkswagen Bus was not as popular in the film industry during the 1980s as in the 1960s and 1970s. However, it did make appearances in some movies during the 1980s, such as *Fast Times at Ridgemont High* (1982) and *Little Miss Sunshine* (2006). The popularity of the Volkswagen Bus as a cultural icon and symbol of the counterculture movement of the 1960s and 1970s may have contributed to its continued use in films and TV shows as a representation of a certain era or lifestyle. The Volkswagen bus has been featured in several famous movies and television shows over the years, including:

- *The Love Bug* – 1968
- *Herbie Goes Bananas* – 1980
- *Fast Times at Ridgemont High* – 1982
- *Forrest Gump* – 1994
- *That 70s Show,* TV series – 1998-2006
- *Austin Powers: The Spy Who Shagged Me* – 1999
- *Scooby-Doo 2: Monsters Unleashed* – 2004

- *Little Miss Sunshine* – 2006
- *Zombieland* – 2009
- *Once Upon a Time in Hollywood* – 2019

The positive impact that Volkswagen had on transportation and mobility cannot be overstated. The company's commitment to quality and innovation has led to the development of some of the most reliable and efficient vehicles in history. Volkswagen's extremely iconic vehicle images are etched in the counterculture and folklore for the ages.

30

Jeep

From military beginnings to most popular outdoor SUV, Jeep transcends ages

John Willys

"The Jeep is the most useful vehicle ever designed."
—General George C. Patton

Jeep is an iconic brand in the automotive industry and has built a rich history of innovation and success. The brand was born from the need for a versatile and rugged vehicle that could handle the demands of military operations during World War II. Today, Jeep is a global brand and is known for its off-road capabilities, rugged design and adventurous spirit. This chapter explores the history of Jeep and its journey as it became a leading automotive brand and the foremost all-terrain vehicle. Jeep ushered in the beginning of the SUV-crazed era of cars.

The origins of Jeep can be traced to the early 1940s when the US Army put out an official request for a specially designed four-wheel-drive vehicle that could handle the rugged terrain of the battlefield. The US Army needed a vehicle that could transport troops, effectively tow artillery to activation sites and perform reconnaissance missions. The new vehicle would need to be lightweight, durable and fully capable of navigating through mud, snow and all forms of rough terrain and landscape.

The task of designing the new vehicle was awarded to three companies: Willys-Overland, Ford Motor Company, and Bantam. The three companies submitted designs, and in the end, the Willys-Overland design was chosen. The Willys-Overland design was based on the company's civilian vehicle, the Civilian Jeep (CJ), which had been in production since the 1930s.

The new military vehicle, which was officially named the Willys MB, went into production in 1941. The Willys MB was a four-wheel-drive vehicle that was powered by a four-cylinder engine and had a top speed of 65 miles per hour. The vehicle was designed to be lightweight and easy to transport, so it could be quickly deployed to the front lines.

The Willys MB was an immediate success and quickly became a favorite among soldiers. The vehicle's rugged design, off-road capabilities and reliability made it ideal for military operations. The Willys MB was used in a variety of roles, including transportation and reconnaissance, and could even serve as an ambulance.

After the war ended, Willys-Overland began producing a civilian version of the Willys MB, which they named the Jeep CJ-2A. The CJ-2A was designed for farmers, ranchers and others who needed a versatile and reliable vehicle for their work. The CJ-2A became a big hit with consumers and quickly became a popular, must-have vehicle for off-road enthusiasts.

In 1950, Willys-Overland introduced the Jeep M38, which was a military version of the CJ-3A. The M38 was used by the US military for implementation during the Korean War and was also sold to foreign governments. In 1953, Willys-Overland was purchased by Kaiser Motors, which continued to produce the Jeep line of vehicles.

During the 1960s, Jeep introduced the Wagoneer, which was a four-wheel-drive vehicle that was designed for families. The Wagoneer was popular with consumers and represented the first four-wheel-drive vehicle to be marketed as a luxury vehicle.

In the 1970s, Jeep introduced the Cherokee, which was a smaller and more fuel-efficient version of the Wagoneer. The Cherokee was also a hit with consumers and quickly became one of the best-selling vehicles in the Jeep lineup.

In 1987, Chrysler Corporation purchased American Motors Corporation, which owned the Jeep brand. Chrysler continued to produce the Jeep lineup of vehicles and introduced new models, including the Grand Cherokee, Wrangler and Liberty.

During the 1990s, Jeep continued to innovate and introduced the first SUV with a unibody design, the Jeep Grand Cherokee. The Grand Cherokee was popular with consumers and became the first SUV to feature a driver's-side airbag.

In 2007, Chrysler was purchased by private equity firm Cerberus Capital Management. Under Cerberus, Jeep continued to produce new models, including the Patriot, Compass and Renegade.

In 2009, Chrysler filed for bankruptcy and was purchased by Italian automaker Fiat. Under Fiat, Jeep continued to produce new models, including the Cherokee and the Wrangler JK.

In 2018, Fiat Chrysler Automobiles (FCA) announced plans to invest $4.5 billion in Jeep over the next five years. The investment is part of FCA's plan to expand the Jeep brand and increase production. FCA planned to introduce new models, including a pickup truck and a three-row SUV.

In 2020, FCA and PSA Group (Peugeot) merged, creating the fourth-largest automaker, worldwide, by volume under the brand Stellantis. All 12 brands remain intact, including Fiat, Dodge, Ram,

Chrysler, Alfa Romeo, Maserti, Mopar (accessories), Peugeot, Citroen, DS, Opel and Vauxhall.

The Jeep marque has been headquartered in Toledo, Ohio, since Willys-Overland launched production of the first Civilian Jeep (CJ) branded models there in 1945. Its replacement, the conceptually consistent Jeep Wrangler series, has remained in production since 1986. With its solid axles and open top, the Wrangler has been called the Jeep model that is as central to the brand's identity as the 911 is to Porsche.

Over the past 14 years, I've relied on the modern Jeep Grand Cherokee as a daily driver. They have all been reliable, stylish, pleasurable to drive and a good fit for Colorado-based operations.

31

Ferrari

Building Ferrari, one of most legendary car brands

Enzo Ferrari

"The Ferrari is a dream—people dream of owning this special vehicle and for most people it will remain a dream apart from those lucky few."
—Enzo Ferrari

Ferrari is a name synonymous with luxury, high-performance cars. The company was founded by Enzo Ferrari, a man who had a passion for racing and a vision for creating the ultimate driving machine. Over the years, the company has grown into a global brand recognized for its high-quality engineering, innovative design and unparalleled performance. In this chapter, we explore the start and growth of the Ferrari car company over the years and delve into the life of Enzo Ferrari, the man behind the brand.

The early years

Enzo Ferrari was born in Modena, Italy, in 1898. He grew up in a family involved in the transportation industry, and he developed a passion for cars at a young age. In 1915, at age 17, Ferrari began working for a local car manufacturer called CMN. He quickly rose through the ranks and became a test driver for the company. However, his dreams of racing were put on hold when he was drafted into the Italian army during World War I.

After the war, Ferrari returned to CMN, but soon grew restless and decided to pursue his passion for racing. In 1920, he joined Alfa Romeo as a racing driver, and he quickly gained a reputation as one of the best drivers in Italy. Ferrari soon realized, though, that he was more interested in the engineering and design of cars than in driving them. In 1929, he left Alfa Romeo and founded Scuderia Ferrari, a racing team that would go on to become one of the most successful in history.

Enzo Ferrari had several unique interests and characteristics, including:

- Passion for racing: Enzo Ferrari was passionate about racing and spent his entire life dedicated to it.

- Attention to detail: Enzo Ferrari was known for his meticulous attention to detail, which was reflected in the design and engineering of his cars. He was involved in every aspect of the car's development, from the engine to the bodywork.

- Business acumen: Enzo Ferrari was a shrewd businessman and was able to turn his passion for racing into a successful business. He was known for his ability to negotiate and close deals, and he built the Ferrari brand into one of the most iconic and recognizable in the world.

- Strong personality: Enzo Ferrari was known for his strong personality and his determination to succeed. He was often described as stubborn and demanding, but also charismatic and inspiring.

- Love for art and culture: Enzo Ferrari was also interested in art and culture, and he was known to have a collection of paintings and sculptures. He was a patron of the arts and supported several cultural initiatives in Italy.

The founding of Scuderia Ferrari

Scuderia Ferrari was initially founded as a racing team that was sponsored by Alfa Romeo. Ferrari soon realized that he wanted to build his own cars, and in 1940 he founded Auto Avio Costruzioni, a company that would later become Ferrari. The company's first car, the 125 S, built in 1947, was an instant success, powered by a V-12 engine designed by Ferrari himself, and it was capable of reaching speeds of up to 150 mph.

Ferrari gained a reputation for building high-performance cars that were designed for racing. The company's early cars were built with a focus on performance, and were often used in races such as the Mille Miglia and the Targa Florio. Ferrari's success was not limited to racing. The company also began to produce road cars designed for the wealthy and elite. These cars were known for their luxurious interiors and high-performance engines, and they quickly became a symbol of status and wealth.

The golden years

The 1950s and 1960s were the golden years for Ferrari. During this time, the company dominated the world of racing, winning countless championships and races. Ferrari's success was due in large part to the genius of Enzo Ferrari, who was known for his meticulous attention to detail and his ability to motivate his team. Ferrari's cars were also known for their advanced engineering, including the use of aerodynamics, lightweight materials and advanced engines.

One of the most famous cars during this time was the Ferrari 250 GTO, built specifically for racing. The 250 GTO won countless races, including the Tour de France and the 24 Hours of Le Mans, and is considered one of the greatest racing cars of all time.

Ford versus Ferrari

The Ford versus Ferrari rivalry is a legendary story in car racing history. It began in the 1960s when Henry Ford II decided to take on Ferrari in the world's most prestigious race, the 24 Hours of Le Mans.

Ford tried to buy Ferrari in 1963, but negotiations fell apart. In response, Ford set out to build a car that could beat Ferrari at Le Mans and hired Carroll Shelby, a former race car driver and designer, to lead the project.

Shelby and his team developed the Ford GT40, a sleek and powerful car that was designed to take on Ferrari. In 1966, Ford won the 24 Hours of Le Mans for the first time, with the GT40 taking the top three spots. The Ford GT40 went on to win the race for the next three years, cementing Ford's dominance over Ferrari.

The rivalry between Ford and Ferrari continued throughout the 1960s and 1970s, with both companies developing new cars and technologies to try and gain an edge. Ford's success at Le Mans marked a turning point in car racing history, as it proved American car companies could compete with the best in the world.

Today, the Ford GT remains an iconic car, and the Ford versus Ferrari rivalry, even immortalized in the 2019 film *Ford vs. Ferrari,* is still remembered as one of the greatest in car racing history.

The decline and later resurgence of Ferrari

The 1970s and 1980s were a difficult time for Ferrari. The oil crisis of the 1970s had a major impact on the company, and it struggled to keep up with the changing demands of the market. In addition, Enzo Ferrari's health began to decline, and he became less involved in the day-to-day operations of the company. During this time, Ferrari's cars began to lose their edge, and the company struggled to compete with other high-performance car manufacturers.

In the 1990s, Ferrari began to experience a resurgence. The company's new CEO, Luca di Montezemolo, was determined to bring Ferrari back to its former glory, and he launched a series of initiatives aimed at revitalizing the company. He invested heavily in research and development, and he worked to improve the quality.

Ferrari, an Italian luxury sports car manufacturer, has undergone significant changes and growth since the 1990s. The company has evolved from a small, niche manufacturer to a global brand with a strong reputation for high-performance vehicles and cutting-edge technology.

In the early 1990s, Ferrari was producing only a few hundred cars a year and was financially struggling to keep up with its competitors and stay relevant in the market. However, in 1997, Ferrari was purchased by the Fiat Group, which provided the financial support and resources necessary for Ferrari to once again grow.

Under the leadership of Montezemolo, Ferrari began to focus on expanding its product line and increasing its global presence. The company introduced the 360 Modena, the F430 and the California, which helped to attract new customers and increase sales.

In addition, Ferrari invested heavily in research and development, particularly in the area of technology. The company introduced several new technologies, such as the F1 gearbox, which allowed for faster gear changes and improved performance. Ferrari also developed a new lightweight chassis, which helped to improve the handling and agility of its vehicles.

Ferrari's growth and success in the 2000s were also fueled by its success in Formula One racing. The company won several championships during this time, which helped to increase its brand recognition and attract new customers.

In recent years, Ferrari has continued to evolve and adapt to changing market conditions. The company has introduced new models such as the 458 Italia and the LaFerrari, which have received critical acclaim for their performance and design. Ferrari has also expanded its global presence, particularly in emerging markets such as China, where the company has opened several new dealerships.

Despite its success, Ferrari has faced several challenges in recent years, including increased competition from other luxury car manufacturers and concerns about environmental sustainability. To address these challenges, Ferrari has invested in hybrid technology and has introduced new models such as the SF90 Stradale, which combines a traditional V-8 engine with electric motors.

Ownership of Ferrari

The history of Ferrari ownership can be traced back to its founding by Enzo Ferrari in 1940. Enzo Ferrari remained the sole owner of the company until 1969, when he sold a 50% stake to Fiat.

Fiat gradually increased its stake in Ferrari over the years, eventually becoming the majority owner in 1988. Yet Enzo Ferrari remained involved with the company until his death in 1988.

In 2014, Fiat Chrysler Automobiles (FCA) announced plans to spin off Ferrari into a separate company. This move was completed in 2016, with Ferrari going public on the New York Stock Exchange under the ticker symbol RACE.

As of 2021, Ferrari is still majority-owned by Exor, the investment company of the Agnelli family, which controlled FCA before the merger with PSA later in 2021.

Ferrari's growth and changes over the past three decades have been impressive. The company has evolved from a struggling niche manufacturer to a global brand with a strong reputation for high-performance vehicles and cutting-edge technology. While it faces challenges in the years ahead, Ferrari's commitment to innovation and excellence suggests that it will continue to thrive in the competitive world of luxury sports car manufacturing.

32

Toyota

From an industrial mill process to the number one automaker in the world – Toyota's story

Kiichiro Toyoda

"Toyota has become a symbol of quality and innovation in the automotive industry. Their cars are known for longevity and dependability."
—*Auto industry analyst*

Toyota Motor Company is a Japanese multinational automaker that is headquartered in Toyota City, Aichi, Japan. Toyota Motor Company was founded by Kiichiro Toyoda in 1937 as a spin-off from his father's company, Toyota Industries. The origins of the company can be traced back to 1924, when Sakichi Toyoda invented the Toyoda Model G Automatic Loom. Toyoda's invention of the Automatic Loom revolutionized the textile industry and made the Toyoda family one of the wealthiest in all of Japan.

Kiichiro Toyoda was interested in the automobile industry and traveled to Europe and the United States to study car production techniques. Toyoda returned to Japan with the goal of creating a Japanese car company that could rival the established American and European manufacturers. In 1933, Toyoda established an automotive department within Toyota Industries and began designing and producing motor vehicles.

The first Toyota car, the Model AA, was introduced in 1936. It was based on the Chrysler Airflow and was powered by a 3.4-liter inline-six engine. The Model AA was well-received in Japan and helped establish Toyota as a serious player in the automotive industry.

The company's growth was interrupted by World War II, during which Toyota produced trucks and military vehicles for the Japanese army. After the war, the company resumed production of cars and began exporting them to other countries.

Edward Deming played a significant role in the adoption and popularization of the principles of quality management and continuous improvement, including Kaizen, within Toyota and the broader Japanese manufacturing industry.

Edward Deming was an American statistician, engineer, and management consultant who became renowned for his expertise in quality control and statistical process control. In the 1950s, Japan was seeking ways to rebuild its economy and improve the quality of its products. Deming's ideas and teachings were instrumental in shaping the quality management practices that Toyota and other Japanese companies adopted.

In 1950, the Union of Japanese Scientists and Engineers (JUSE) invited Deming to Japan to deliver a series of lectures on statistical process control and quality management. His teachings emphasized the importance of reducing variation, improving processes, and creating a culture of continuous improvement. Deming emphasized that quality was not just the responsibility of the inspection department but should be built into every aspect of the production process.

Deming's ideas resonated with Japanese manufacturers, including Toyota. Taiichi Ohno, one of the key architects of Toyota's production system, was heavily influenced by Deming's teachings. Ohno recognized the value of statistical control and the need for continuous improvement to achieve superior quality and efficiency.

Toyota embraced Deming's philosophy and integrated it into the Toyota Production System. Deming's emphasis on empowering employees, eliminating waste, and continually improving processes aligned with Toyota's goal of achieving excellence in manufacturing.

Deming's teachings also had a broader impact on the Japanese manufacturing industry as a whole. The principles of quality management and continuous improvement propagated by Deming were embraced by numerous companies in Japan, leading to a significant shift in the country's manufacturing practices and an enhanced reputation for producing high-quality products.

One of the key people involved in the growth of Toyota was Eiji Toyoda, Kiichiro Toyoda's cousin. Eiji Toyoda joined the company in 1936 and became president in 1967. He was instrumental in establishing Toyota's famous Toyota Production System, which focused on reducing waste and increasing efficiency in manufacturing. This system became a model for other companies and is still used by Toyota today.

In the 1960s, Toyota began exporting cars to the United States. The company's first American dealership was established in 1957, but it wasn't until the 1960s that Toyota began to gain a foothold in the American market. The company's small, fuel-efficient cars were well-suited to the American market during the oil crisis of the 1970s, and Toyota's reputation for quality and reliability helped it gain a loyal following.

A Toyota story that is etched in my memory and will be forever, has to do with when Toyota was first entering US market with their cars, in the early 1960s. Toyota had dealer development people go from city to city and town to town to scout the best prospects for dealers to represent the company in local markets. One stop in a small town in western Colorado had the Toyota dealer development official stop by a Chrysler dealership. The Toyota rep called on the dealership owner and said "You come highly recommended to us to represent Toyota in the market." The third-generation dealer didn't know Toyota cars very well at all. He had never even driven one yet. The dealer asked what it would take to become a dealer. The response was "You would need to sign our sales and service agreement and agree to buy a Toyota sign for your dealership." The dealer asked how much the sign would cost. The answer, "$5,000." The dealer, thinking that was an outrageous price for a sign, rejected the offer to become a Toyota dealer. A competing dealership down the street, a second-generation General Motors dealership owner, agreed to buy the sign and execute the Toyota dealership sales and service agreement.

A few weeks later, a representative from Subaru was in the same town, scouting dealer owners to sign on to become a Subaru dealer in that town. Subaru approached the same Chrysler dealer owner that Toyota recently had approached. Subaru had a similar offer, though the price for the sign was only $2500 versus the much higher $5,000 for the Toyota sign. That dealership is still a Subaru dealer today nearly 70 years and another owner generation later. The other dealer is still a Toyota dealer today, nearly 70 years later. Dealership development for Toyota played out in a similar fashion to that all over the United States in the early 1960s and beyond. Toyota was able to build one of the strongest dealer networks in the country.

In the 1980s and 1990s, Toyota continued to expand its operations around the world. The company established manufacturing plants in the United States, Canada and Europe, and began producing cars specifically designed for these markets. In 1997, Toyota became the first Japanese car company to sell more than 1 million cars in the United States in a single year.

Today, Toyota is one of the largest car companies in the world, with operations in more than 150 countries. The company is known for its innovative designs, focus on quality and reliability, and commitment to sustainability. Toyota has also been a leader in the development of hybrid and electric cars and has set a goal to eliminate carbon emissions from its vehicles by 2050.

The models Toyota builds and sells that have proved most popular in their new vehicle categories are: Camry, 4Runner, Highlander, Rav4, Sienna, Sequoia, Tundra, Tacoma, Prius and the all new Crown. For decades, the Toyota Camry was the fastest selling car in the US and the world. Early in 2023, the Tesla Model 3 ended that long-term, four-decade trend.

Toyota is the hands-down leader in hybrid powertrain technology, in some years selling more hybrid vehicles than all other automakers combined. Now Toyota is moving to bring hybrid and plug-in hybrid (PHEV) technology to all vehicle models. Though leaders in hybrid and PHEV tech, Toyota has been widely criticized for lack of production in pure battery electric vehicle (BEV) technology. Recently it was announced that Toyota will re-structure its BEV production methods and plans to better compete in those markets.

Toyota Motor Company is a success story that began with the vision of Kiichiro Toyoda and the ingenuity of Sakichi Toyoda. The company's commitment to quality, efficiency and innovation has helped it become one of the largest car companies in the world. Eiji Toyoda's contributions to the development of the Toyota Production System were instrumental to the company's growth and success. Toyota's expansion into the United States and other markets around the world helped it become a global automotive behemoth that has been recognized for its quality, durability and reliability. Today, Toyota continues to innovate and lead the way in the automotive industry, while also striving to be a responsible corporate citizen.

33

Kia

Driving one of fastest quality improvement curves in the industry

Chung Ju Yung

"Kia has shown a remarkable improvement in quality and craftsmanship and it's reflected in the satisfaction of their customers."
—*JD Power Automotive*

Kia Motors is a South Korean automobile manufacturer that has grown to become one of the most successful car brands in the world. According to the company, the name "Kia" derives from the Sino-Korean characters 'to arise' and which stands for 'Asia'. The name Kia is roughly translated as "Rising from East Asia."

Since its start in 1944, Kia has gone through a tumultuous journey that has seen it experience both highs and lows. However,

Kia's ability to adapt to changing market conditions and consumer needs has been instrumental, even key, to its success. Today, Kia is a global brand that produces high-quality and reliable cars that are loved by millions of people worldwide. This chapter will cover the history of Kia Motors, its successes, and the meaningful factors that have contributed to its growth and success.

Kia Motors was founded in December 1944 as Kyungsung Precision Industry, a manufacturer of steel tubing and bicycle parts. The company changed its name to Kia Industries in 1952 and began producing trucks and motorcycles. However, it was not until 1974 that Kia began producing passenger cars, with the launch of the Brisa model. The Brisa was a small, four-door sedan that was powered by a 1.4-liter engine. Although the Brisa was not a commercial success, it marked the beginning of Kia's journey into the passenger car market.

In the 1980s, Kia began producing cars under license from foreign automakers including Peugeot and Mazda. This allowed the company to gain valuable experience in car manufacturing and engineering. By the end of the decade, Kia had established itself as a leading car maker in South Korea, with a range of popular models such as the Pride, Sephia and Sportage.

However, the 1990s were a challenging period for Kia. The company faced financial difficulties due to the Asian financial crisis, which hit the region in 1997. As a result, Kia was forced to sell a controlling stake to Hyundai Motor Company in 1998. The acquisition by Hyundai was a turning point for Kia, as it provided the company with the financial resources and expertise needed to compete on a global scale.

Under the ownership of Hyundai, Kia underwent a major transformation. The company invested heavily in research and development, as well as design and marketing, which helped it to improve the quality and reliability of its cars. In 2004, Kia launched the Optima sedan, which was the first car to feature the company's new design language, "Tiger Nose." The bold and distinctive design of the Optima was a departure from the bland and uninspiring designs of Kia's earlier models and helped to establish the brand as a serious player in the global car market.

Kia's success continued in the 2010s, with the company launching a range of new models that were well-received by consumers and critics alike. In 2011, Kia unveiled the Rio hatchback, which was praised for its stylish design, fuel efficiency and affordability. The same year, the company launched the Optima Hybrid, which was the first hybrid car, along with the Hyundai Sonata, to be produced by a Korean automaker. The Optima Hybrid was a significant achievement for Kia, as it demonstrated the company's commitment to developing eco-friendly cars.

In 2014, Kia launched the Soul EV, which was the company's first all-electric car. The Soul EV was a game-changer for Kia, as it demonstrated the company's ability to produce electric cars that were practical, affordable and fun to drive. The Soul EV was praised for its range, which was among the best in its class, and its stylish and practical design.

Today, Kia is a global brand that produces a wide range of cars, from small hatchbacks to luxury sedans and SUVs. The company has a strong presence in markets such as the United States, Europe and China, and is known for producing high-quality and reliable cars that offer excellent value for money. Kia's success can be attributed to a number of factors, including its focus on design, innovation and customer satisfaction.

One of the key factors that has contributed to Kia's success is its focus on design. In recent years, Kia has become known for producing cars that are stylish, distinctive and visually appealing. The company's design language, "Tiger Nose," has become a hallmark of Kia's cars, and is instantly recognizable to car aficionados and enthusiasts around the world. Kia's commitment to design has helped it to stand out in a crowded market and has contributed to its success in attracting younger, more style-conscious consumers.

Another factor that has contributed to Kia's success is its focus on innovation. Kia has invested heavily in research and development and has been at the forefront of developing new technologies such as hybrid and electric cars. The company's commitment to innovation has helped it to stay ahead of the curve and to anticipate changing consumer needs and preferences.

Kia's success can also be attributed to its focus on customer satisfaction. Kia has a reputation for producing cars that are reliable, affordable, and easy to maintain. The company has also invested heavily in customer service, with a range of programs and initiatives designed to enhance the ownership experience for Kia customers. This focus on customer satisfaction has helped Kia to build a loyal customer base and to establish itself as a trusted and respected brand in the global car market.

In 2022, Kia was awarded one of the most JD Power Automotive Awards, including those for quality, performance and reliability. New car dealers with the Kia franchise have done very well with it over the past two to three years. This is a very different situation from just a few years ago when I remember a Kia dealer who was unhappy with the quality of Kia brand cars and frequently called them 'shitboxes'. Its amazing how fast a car brand can move from being an also ran to one of the most coveted franchises to represent in the new car market. Kia is a Cinderella story when it comes to quality, value, reputation and brand image. Every new car brand should be so fortunate.

During 2023, Kia made a mark on electrification when it comes to its brand line of distinctive cars. Aside from the Soul EV, Kia has done very well with a high volume of Niro models (pure EV, hybrid and gasoline-powered versions), as well as a stylish new Kia EV6 which sports among the highest ranges in the industry and offers the fastest charging curve of any modern BEV in the market. Both the Niro EV and the Kia EV6 are solid, reliable cars for the newish electric car market. I've owned one of each, and I'm currently with the EV6. It is one my favorite cars of all time.

34

Honda

The Japanese motorcycle and car company goes global

Soichura Honda

*"The Honda brand is known for its durable, well-built cars
that hold their value over time."*
—Consumer Reports

The story of the creation and founding of Honda Motor Company is one of perseverance, innovation and hard work. The company's journey began in the aftermath of World War II, when Japan was struggling to rebuild its economy. Despite the challenges, a young engineer named Soichiro Honda had a vision of creating a company that would manufacture high-quality, long-lasting motorcycles for the Japanese market. Soichiro Honda was born in Iwata District, Shizyuoka, near Hamamatsu on November 17, 1906.

Honda spent his early childhood helping his father, Gihei Honda, a blacksmith, with his bicycle repair business. Honda's recollections of his youth mostly centered on the family's poverty. Five of his eight brothers and sisters died in childhood.

He went to school for only 10 years, and at age 15 struck out for Tokyo, eager to become a mechanic. He quickly got a job at an auto shop, only to discover that his only responsibility was babysitting. But the next year a tremendous earthquake destroyed Tokyo and Honda got his break: he filled in for mechanics who left their jobs to rebuild their homes.

Honda's early years were marked by setbacks and failures. He started his first company, the Honda Technical Research Institute, in 1946, but it was forced to close just two years later due to lack of funding. Undeterred, Honda continued to develop his ideas, and in 1948 he founded the Honda Motor Company.

In the early years, Honda focused on producing small motorcycles that were reliable and affordable. His first model, the Honda Cub, was an instant success, and it quickly became the best-selling motorcycle in Japan. Over the next few years, Honda expanded the product line to include larger motorcycles and even scooters. The Honda Super Cub has been in continuous manufacture since 1958 with production surpassing 60 million in 2008, 87 million in 2014, and 100 million in 2017. The Super Cub is the most produced motor vehicle in history.

One of the key factors in Soichiro Honda's success was his commitment to innovation. He was constantly experimenting with new designs and technologies, and he was always looking for ways to make his products better. For example, in the 1960s, Honda introduced the first mass-produced four-stroke motorcycle engine, which was more fuel-efficient and environmentally friendly than the two-stroke engines that were common at the time.

Another important factor in Honda's success was his focus on quality. Honda believed that if he could design products that were better than his competitors, customers would be willing to pay a premium for them. To achieve this, Honda implemented a rigorous quality control system that ensured that every product that left the factory met his high quality standards.

As business grew at Honda, he began to look beyond the Japanese market. In the 1960s, he began exporting his motorcycles to other countries, including the United States. The first Honda motorcycle dealership in the US was granted in 1959. At first, Honda faced stiff competition from established American motorcycle manufacturers, but his products quickly gained a reputation for reliability and quality.

One of the most popular Honda models in the United States was the Honda CB750, which was introduced in 1969. The CB750 was a groundbreaking motorcycle that featured a powerful four-cylinder engine, disc brakes and other advanced features that were not commonly found on motorcycles at the time. The CB750 was an instant hit, and it helped establish Honda as a major player in the American motorcycle market.

In the following years, Honda continued to expand its product line to include cars, ATVs and other types of vehicles. One of the most popular Honda cars in the United States was the Honda Civic, which was introduced in 1972. The Civic was a compact car that was fuel-efficient, reliable and affordable and it quickly became a favorite among American drivers.

Today, Honda is one of the largest and most successful automotive manufacturers in the world. It employs over 200,000 people and has manufacturing facilities in countries around the globe. Honda's success can be attributed to the vision and hard work of its founders, as well as the company's commitment to innovation, quality and customer satisfaction.

In 1989, Soichiro Honda was inducted into the Automotive Hall of Fame near Detroit.

Soichiro Honda's vision of creating a company that would manufacture high-quality motorcycles for the Japanese market has evolved over the years to include a wide range of products. The key to Honda's success has been its commitment to innovation, quality, and customer satisfaction. With a strong global presence and a reputation for reliability and quality, Honda is poised to continue its success for many years to come.

35

Land Rover

The popular story of a high-level European automotive brand

Maurice Wilks

"The Land Rover has a unique place in the automotive world. It's the only vehicle that can truly go anywhere."
—Sir William Lyons, co-founder of Land Rover

Land Rover is one of the most sought-after automotive brands in the world. Founded in 1948, the company has a deep and very rich history of innovation, adventure and durability that has made it a favorite among drivers and auto enthusiasts alike. From Land Rover's earliest days as a utilitarian off-roader to its current position as a luxury SUV brand, it has always been at the forefront of automotive technology and design.

In this chapter we explore the founding and creation of Land Rover as an automotive brand and company and its growth and successes through the years.

The founding of Land Rover

Land Rover was founded in 1948 by the Rover Company, a British-based manufacturer of luxury cars. The idea for the company came from Rover's highly talented chief designer, Maurice Wilks, who was inspired by the American Jeep he had seen on a beach in Wales.

Wilks saw the potential for a British-made off-road vehicle that could be used by farmers as well as other rural workers. Wilkes designed a prototype on his farm in Anglesey, using an old Jeep chassis and incorporating a Rover engine. The result was a rugged, all-terrain vehicle that could handle any type of landscape and under all types of weather conditions.

The first-ever Land Rover was unveiled at the Amsterdam Motor Show in 1948 and it was an instant success. The company received orders for hundreds of vehicles and began production in earnest at Rover's Solihull factory in Birmingham.

The early Land Rovers were simple, utilitarian vehicles designed for work rather than pleasure. They had a canvas roof and a basic interior with metal seats. But they were incredibly rugged and reliable. Land Rovers quickly gained a great reputation for being able to go anywhere and do anything.

The growth of Land Rover

Over the years, Land Rover continued to grow and evolve. In the 1950s, the company introduced a number of new models and styles, including the Series II and III, which were more comfortable and refined than the original Land Rover. These vehicles were still primarily used for work but they also began to attract interest from recreational drivers who appreciated their off-road capabilities.

In the 1960s, Land Rover introduced the elaborate and high-end Range Rover, a luxury off-road vehicle that was designed to compete with the likes of the Jeep Wagoneer and the Ford Bronco. The Range Rover was a huge success and it quickly became one of the most popular and sought after SUVs in the world.

Over the years, Land Rover continued to innovate and improve its vehicles in quality, which was lacking in the early years. In the 1970s, the company introduced the Series III Lightweight, a stripped-down version of the Land Rover that was designed for military use. In the 1980s, Land Rover introduced the Discovery, a mid-size SUV that was designed for families and recreational drivers.

Evolution of ownership for Land Rover, over the years

In the 1990s, Land Rover was acquired by BMW, which invested heavily in the company and helped to modernize its production facilities. BMW also introduced a number of new models, including the Freelander, a compact SUV that was designed for urban drivers.

In 2000, Land Rover was sold to Ford, which continued to invest in the company and expand its product line. Ford introduced the Range Rover Sport, a high-performance version of the Range Rover, and the LR3, a mid-size SUV that was designed for off-road enthusiasts.

In 2008, Land Rover was sold to Tata Motors, an automaker headquartered in India, which has continued to invest in the company and expand its product line. Tata Motors introduced the Range Rover Evoque, a compact SUV that was designed for urban drivers, and the Discovery Sport, a mid-size SUV that was designed for families and recreational drivers.

The success of Land Rover

Over the years, Land Rover has become one of the most successful and reliably profitable automotive brands in the world. The company has won numerous awards for its vehicles, including the prestigious Car of the Year award from *Motor Trend* magazine.

Land Rover's success can be attributed to a number of factors. First and foremost, the company has always been at the forefront of automotive technology and design. Land Rover has consistently introduced new features and innovations that have helped to set it apart from its competitors.

Secondly, Land Rover has always been committed to quality and durability, even more so in recent years. Today, the company's vehicles are built to last, and they are designed to handle even the toughest terrain and weather conditions.

Finally, Land Rover has always been associated with adventure and exploration. The company's vehicles are designed to take drivers off road for excursions and go anywhere the driver wants to go. They largely will, with some limitations, of course.

Jaguar – Land Rover made a conscious decision in 2022 to move upstream with their Jaguar brand cars, and in the process to reduce the worldwide dealer count. Land Rover was already higher end and did not reduce dealer count, though they did increase expectations for the dealer body already in place. Almost all Land Rover dealers, like the parent company, are profitable and successful. Dealers just hope to increase allotment and be able to gain more sales based on higher levels of production.

36

Subaru

Establishes itself as versatile and utilitarian for its popularity

Kenji Keti

"Subaru has built a reputation for producing vehicles that are safe, reliable and capable in all kinds of weather conditions."
—*Car and Driver*

Subaru is a Japan-based auto manufacturer that has come a long way since its inception in 1953. The company started as Fuji Heavy Industries, a conglomerate that manufactured airplane parts for the Japanese military. However, after the war, the company shifted its focus to manufacturing motor vehicles. The Subaru brand was officially launched in 1955, and since then it has become one of the most successful auto manufacturers in the world.

The success of Subaru can be attributed to its focus on innovation and quality. Subaru has been at the forefront of technology, and it was the first Japanese automaker to introduce all-wheel drive capabilities in its cars. This technology has given Subaru a significant edge over its competitors and helped it establish a reputation for producing highly versatile, go anywhere, reliable vehicles.

One of the most famous Subaru models is the Impreza, which was introduced in 1992. The Impreza quickly became popular among enthusiasts due to its rally-inspired design and impressive performance. It was also the vehicle that introduced the world to the legendary Subaru WRX, which is still considered one of the best and most affordable performance cars in the market today.

Another very iconic Subaru model is the Outback, introduced in 1994. The Outback was designed to be a rugged, off-road vehicle that could handle most any terrain. The Outback quickly became popular among outdoor enthusiasts and families due to its spacious interior and all-wheel drive capabilities.

In recent years, Subaru has continued to innovate and expand its lineup of vehicles. The company has introduced new models, such as the Ascent, a three-row SUV, and the Crosstrek, a smaller crossover. These models have helped Subaru appeal to a wider range of customers and continue to grow its market share.

Subaru's success in the automotive industry has also been due to its commitment to safety. It was one of the first automakers to feature antilock brakes and airbags. Today, all Subaru models come with advanced safety features such as EyeSight Driver Assist Technology, which uses cameras and sensors to monitor the road and help prevent accidents.

In some specific states and regions in the United States, Subaru has moved from its position as 8th, 9th or 10th in volume sales as an automaker nationally, up the sales ladder to as high as 3rd or 4th in volume sales. States that Subaru outperforms its national position include but are not limited to: Colorado, Vermont, Montana, New Hampshire, and Maine. For example, in Colorado it is fairly normal for Subaru to rank as 3rd highest volume in new car sales, behind Toyota and Ford. At least one month in 2017, Subaru actually moved to the #1 volume sales position in Colorado, above all other automakers. It was a rare and short-lived feat, though still extremely impressive as a brand that performs much lower overall nationally.

The Atlantic ran a story in 2016 about Subaru's, at the time, gutsy move to target lesbians as a market segment. Here are excerpts from that article.

In the 1990s, Subaru's unique selling point was that the company increasingly made all-wheel drive standard on all its cars. When the company's marketers went searching for people willing to pay a premium for all-wheel drive, they identified four core groups who were responsible for half of the company's American sales: teachers and educators, health-care professionals, IT professionals, and outdoorsy types.

Then they discovered a fifth: lesbians. "When we did the research, we found pockets of the country like Northampton, Massachusetts, and Portland, Oregon, where the head of the household would be a single person—and often a woman," says Tim Bennett, who was the company's director of advertising at the time. When marketers talked to these customers, they realized these women buying Subarus were lesbian.

"There was such an alignment of feeling, like [Subaru cars] fit with what they did," says Paul Poux, who later conducted focus groups for Subaru. The marketers found that lesbian Subaru owners liked that the cars were good for outdoor trips, and that they were good for hauling stuff without being as large as a truck or SUV. "They felt it fit them and wasn't too flashy," says Poux.

Subaru's strategy called for targeting these five core groups and creating ads based on its appeal to each. For medical professionals, it was that a Subaru with all-wheel drive could get them to the hospital in any weather conditions. For rugged individualists, it was that a Subaru could handle dirt roads and haul gear. For lesbians, it was that a Subaru fit their active, low-key lifestyle.

Although it was easier to get senior management on board with making ads for hikers than for lesbians, the company went ahead with the campaign anyway. It was such an unusual decision—and such a success—that it helped push gay and lesbian advertising from the fringes to the mainstream. People joke about lesbians' affinity for Subarus, but what's often forgotten is that Subaru actively decided to cultivate its image as a car for lesbians.

Subaru is a company that has come a long way since its early days as a manufacturer of airplane parts. Its focus on innovation, quality and safety has helped it become one of the most successful automotive brands in the world. The Impreza, Outback, Crosstrek and Impreza are amazing Subaru models that have helped establish the brand's reputation for producing high-performance, reliable vehicles. With its commitment to innovation and safety, and all-wheel drive reputation, Subaru is sure to continue its success for years to come.

37

AMC

The rise and fall of American Motors Corporation

Gerald Meyers

American Motors has been a leader in the automotive industry for many years, and we are proud to have contributed to the success of the company."
—*George Romney, former governor of the State of Michigan and former president of American Motors Corporation*

The American Motors Corporation (AMC) was created in 1954 through the merger of Nash-Kelvinator Corporation and Hudson Motor Car Company. The idea behind the merger was to create a company that could better compete with the "Big Three" automakers of General Motors, Ford, and Chrysler. However, it was not until the appointment of George Romney as CEO in late 1954 that AMC began to gain traction and make a name for itself in the industry.

George Romney was born in 1907 in Chihuahua, Mexico, to American parents who were living there as Mormon missionaries. He grew up a devout Mormon in Idaho and Utah and later attended Brigham Young University. In 1926, Romney dropped out of college to serve a two-year mission in Great Britain for the Church of Jesus Christ of Latter-day Saints. Upon his return, he worked a variety of jobs, including as a salesman for Alcoa, a manufacturer of aluminum products.

In 1939, Romney was hired by the automobile manufacturer Nash-Kelvinator Corporation as a manager of their Detroit plant. He quickly rose through the ranks and was appointed president/CEO of the company in 1954, just as the merger with Hudson was taking place. Romney was a dynamic and energetic leader who was committed to making AMC a success.

One of the first things that Romney did as CEO of AMC was to restructure the company's management team. He brought in new talent and created a more streamlined and efficient organization. He also implemented a number of cost-cutting measures including eliminating a number of unprofitable models that had arrived with the merger of the two companies.

Romney was also a strong advocate for innovation and new technology. Under his leadership, AMC developed a number of groundbreaking vehicles, including the Rambler, which was the first compact car produced by an American automaker. The Rambler was a huge success and helped to establish AMC as a serious player in the industry.

In addition to his focus on innovation and efficiency, Romney was also known for his commitment to corporate social responsibility. He believed that businesses had a responsibility to their employees, customers, and communities and he strived to ensure that AMC was a good corporate citizen. He implemented a number of programs to improve working conditions and provide benefits to employees, such as health insurance and retirement plans. He also established a scholarship program for employees' children and supported a large number of charitable organizations.

Romney's leadership style was characterized by his charisma, energy, and optimism. He was a master at effective public relations and was known for his ability to connect with people. Romney was also a skilled negotiator and was able to secure favorable deals with suppliers and labor unions.

Despite his many successes, Romney's tenure at AMC was not without its challenges. In the late 1950s and early 1960s, the company faced intense competition from the "Big Three" automakers, who were producing larger and more powerful cars. In response, AMC tried to compete by producing larger cars of their own, but this strategy proved unsuccessful.

Another challenge that Romney faced was the rise of foreign competition, particularly from newly arriving Japanese automakers. In the 1960s, AMC began to import cars from Japan and Europe in an effort to compete with these new players in the market.

Despite these challenges, Romney remained committed to his vision for AMC. He continued to innovate and introduce new models, such as the Javelin and the AMX, which were designed to appeal to younger, performance-oriented drivers. He also continued to invest in new technology, such as the development of an early version of the hybrid engine.

In 1962, Romney left AMC to pursue a career in politics. Later that year Romney was elected governor of Michigan and he later ran for president of the United States in 1968. Although he was unsuccessful in his presidential bid, his legacy at AMC lived on. The company continued to produce innovative and popular vehicles, such as the Jeep Cherokee and the Eagle and was eventually acquired by Chrysler in 1987.

George Romney's tenure as CEO of AMC was a period of great innovation and success for the company. Through his leadership and vision, AMC was able to establish itself as a serious player in the thriving automobile industry and produce a number of groundbreaking vehicles. Romney's commitment to corporate social responsibility and his ability to connect with people also helped make AMC a respected and admired company. Despite facing a number of challenges, Romney remained committed to his vision for AMC and his legacy continues to influence the automotive industry still today.

There is a legendary story about the design of the AMC Gremlin, an American subcompact car produced by AMC from 1970 to 1978.

The story goes that in the mid-1960s, AMC's CEO at the time, Roy D. Chapin Jr., wanted to develop a small car that would compete with the growing popularity of imported compact cars, particularly those from Europe and Japan. He challenged AMC's design team to come up with a new car that could be built quickly and cost-effectively.

The idea for the Gremlin began in 1966 when design chief at AMC, Dick Teague, and stylist Bob Nixon discussed the possibility of a shortened version of AMC's compact car. On an airline flight, Teague's solution, which he said he sketched on an air sickness bag, was to truncate the tail of a Javelin. Bob Nixon joined AMC as a 23-year-old and did the first formal design sketches in 1967 for the car that was to be the Gremlin.

Dick Teague gathered his team to create a compact car that would be economical, practical, and have a unique appearance. The team was given only a few months to develop the new car, which put them under significant pressure.

During the design process, Teague and his team faced some limitations, including a short development timeframe and the need to utilize existing components and platforms from other AMC models. Instead of the Javelin they decided to use the platform of the AMC Hornet, a larger car, and modify it to create a smaller, more compact vehicle.

The most distinctive aspect of the Gremlin's design came about because of the engineers shortening the Hornet's wheelbase by 12 inches (305 mm). This modification left a significant amount of space behind the rear wheels, which the designers decided to retain rather than attempt to fill in or extend the car's length.

The decision to keep the car's unique rear design, with a truncated back end and a large glass hatch, resulted in the car's distinctive appearance. Some have described it as having a "chopped-off" or "squashed" look, while others found it quirky and endearing.

Despite its unconventional appearance, the Gremlin became a commercial success for AMC, especially during the fuel crisis of the 1970s when smaller, more fuel-efficient cars were in high demand. The Gremlin's compact size, affordable price, and distinctive design made it popular among consumers.

So, the story of the AMC Gremlin's design is one of ingenuity and resourcefulness, with the design team making the most of their available resources and time constraints to create unique and successful vehicles.

38

Mini

The Mini invasion of the US and the world

Sir Alec Issigonis

*"The Mini is a car that has transcended generations and continues
to be a symbol of style and innovation. From its role in the
swinging '60s to its modern-day revival, the Mini has
cemented its place in pop culture as an enduring icon."*
—*Sir Alec Issigonis, designer of the original Mini*

The Mini is a British automotive marquis brand. It was founded in
1959 by the British Motor Corporation (BMC), which evolved
out of the Cooper Car Company. The Cooper Car Company was
named for its founder, John Cooper. Of the Mini, John Cooper said,
"The Mini is not just a car. It's a personality statement. It's a symbol of
freedom, fun and individuality." The Mini quickly became an iconic
symbol in British culture and design and it has since become one

of the most successful and beloved small cars in the automotive industry'S history.

Sir Leonard Lord of the Morris Company issued his top engineer, Sir Alec Issigonis, a challenge: design and build a small, fuel-efficient car capable of carrying four adults, within economic reach of just about everyone. As fate would have it, the challenge of fitting so much function into such a small package inspired a couple of historic innovations. The two most important innovations Issigonis came up with were to create more room in the cockpit: pushing the wheels all the way out to the corners and turning the engine sideways giving the car more stability in tight turnS and more passenger space on the inside. The world had never seen a car quite like it. And when the first Mini launched in 1959, the public was a bit baffled.

Soon enough, people began to recognize that the Mini was not merely a car. The unique combination of classic British style in a low-cost, small size, fun and nimble package came to symbolize independence and spontaneity: the very essence of the youthful 1960s culture. One of the most remarkable elements of the Classic Mini's popularity was how its infectious spirit transcended traditional class barriers. From hipsters and mods to milkmen, rock stars and royalty to rally racers. Everyone could have efficiency, fun and freedom, motoring in a Mini.

The Mini was an instant success upon its launch, winning numerous awards and accolades for its innovative design and exceptional performance. The Mini was featured in numerous famous movies and television shows adding to its cultish popularity.

One of the most notable appearances of the Mini was in the 1969 classic film, *The Italian Job*. The film featured a group of thieves who used three Mini Coopers to carry out their heist. The Minis were showcased as the perfect vehicles for navigating through narrow streets and tight corners, making them an essential part of the heist.

The Mini has also been featured in popular television shows such as *Mr. Bean,* where the titular character, played by Rowan Atkinson, drove a bright yellow Mini. The car became a defining feature of the show and was used in various comedic situations, including being driven from the roof and being crushed by a tank.

Another popular television show that featured the Mini was *The Saint,* which aired from 1962 to 1969. The show starred Roger Moore as Simon Templar, a master thief who drove a white Mini Cooper. The car was often used in high-speed chases and became an iconic part of the show.

In recent years, the Mini has continued to make appearances in movies and television shows. In the 2003 remake of *The Italian Job,* the Mini was once again featured as a key part of the heist. The car was also prominently featured in the popular British television show, *Top Gear,* where it was put through its paces in various challenges and races.

In the famous Austin Powers movie *The Spy Who Shagged Me,* a Mini was featured as part of its lineup. The Mini was famously highlighted in the film's opening scene, where Austin Powers drove the car through the streets of London in a high-speed chase. The Mini car has proven to be a versatile and enduring vehicle that captures the imagination of audiences around the world.

Over the years the Mini brand has undergone several changes in ownership and management, yet it has remained a popular and enduring symbol of British style. During the 1980s, Mini was acquired by British Aerospace and in the 1990s it was sold to German automaker BMW.

Under BMW's leadership, Mini has experienced a period of significant growth and success. BMW invested heavily in the brand, introducing several new models and expanding the global reach of Mini. Today the Mini is sold in over 100 countries worldwide and has become one of the most popular brands in the small car market.

One of the keys to the Mini's success has been its ability to evolve and adapt to changing consumer taste and preferences. While the brand's classic design and heritage remain a central part of its identity, Mini has been adept at embracing new technologies and innovations to keep pace with the changing automotive landscape.

In recent years, Mini has introduced several new models, including the Mini Countryman SUV, the Mini Paceman crossover and the Clubman, a station wagon type vehicle. These new models have helped to broaden the brand appeal for Mini and attract new customers who may not have considered a Mini in the past.

Overall, Mini has a rich and storied history of success and innovation. From its revolutionary design in the 1950's to its current position as a global automotive brand, Mini has remained a beloved and iconic new car brand. Its earliest electric vehicle (EV) models, though not necessarily long in range, were a popular hit with consumers. BMW's positive track record building high quality and popular electric vehicles, bodes well for Mini's prospects in the EV space for the future. Mini has announced it plans to go electric with all of its vehicle models by 2030.

39

Mazda

A journey of innovation and success

Jujior Matsuda

*"Mazda is a company that has always been driven by the
spirit of innovation and the desire to create cars
that deliver a unique driving experience."
—Takashi Yamanouchi, former president
and CEO of Mazda Motor Corporation*

The Mazda Car Company has become a prominent player in the automotive industry and is most known for its innovative designs and high-performance vehicles.

The founding of Mazda

The history of Mazda goes back to 1920, when Jujiro Matsuda established Toyo Cork Kogyo Co., Ltd., a company that initially produced cork products. Matsuda had a vision to innovate and expand the company's horizons. In 1931, the company began manufacturing three-wheeled trucks known as "Mazda-Go." It was during this time that the company adopted the name "Mazda," derived from the Zoroastrian god of wisdom, Ahura Mazda, which is intended to symbolize the company's commitment to intelligence and creativity.

Mazda's key leaders

The key leader in the founding of Mazda was Matsuda, who served as the company's first president. Matsuda was a visionary entrepreneur who believed in the power of innovation. He laid the foundation for Mazda's success by fostering a culture of creativity and encouraging his employees to think outside the box.

Another influential figure in Mazda's history was Kenichi Yamamoto, who joined the company in 1945 as an engineer. Yamamoto played a crucial role in developing Mazda's signature rotary engine, which became a defining feature of the company's vehicles. His leadership and technical expertise propelled Mazda's success in the 1970s and 1980s.

Early success of Mazda

Mazda's early success can be attributed to its commitment to innovation and the introduction of groundbreaking technologies. In 1960, the company unveiled its first passenger car, the Mazda R360 Coupe. This compact and affordable vehicle quickly gained popularity in Japan, becoming the best-selling car in the country at the time. The success of the R360 Coupe set the stage for Mazda's expansion into the global market.

One of the pivotal moments in Mazda's history came in 1967 with the introduction of the Cosmo Sport, the world's first production car powered by a rotary engine. This breakthrough technology showcased Mazda's engineering prowess and positioned the company as a pioneer in the automotive industry. The rotary engine became synonymous with Mazda and was featured in many of its most recognized models, including the RX-7 and RX-8.

Under the leadership of Yamamoto, Mazda experienced significant growth in the 1970s and 1980s. The company's commitment to performance and innovation was highlighted by the launch of the Mazda MX-5 Miata in 1989, a two-seat sports car which captured the hearts of enthusiasts worldwide and remains one of Mazda's most iconic models to this day.

The Mazda MX-5 Miata is, in my opinion, the most popular two-seat convertible coupe and has been for over a generation. It is the space Mazda owns in automotive culture.

The Mazda rotary engine by Wankel

Mazda's interest in the rotary engine began when they acquired the rights to produce Wankel engines in the early 1960s. Felix Wankel, a German engineer, had invented the rotary engine, which offered a more compact and lightweight design compared to conventional piston engines. The concept involved a triangular rotor moving in an oval chamber, creating a continuous cycle of compression, ignition, and exhaust.

However, the rotary engine was uncharted territory, and other automakers had struggled with its complex engineering challenges. Mazda's engineers were undeterred and fully committed to making the rotary engine work. Led by Kenichi Yamamoto, a brilliant engineer who later became Mazda's president, they embarked on an ambitious journey to develop a reliable, mass-produced rotary engine.

The development process was arduous and involved countless hours of research, testing, and fine-tuning. The engineers faced numerous technical difficulties, including sealing issues, combustion inefficiencies, and oil consumption problems. They had to overcome each obstacle step by step, constantly improving the design and materials to make the engine more durable and efficient.

Mazda's 24 hours of Le Mans

Since the company's founding, it has relentlessly taken on new challenges. Symbolic of this "never stop challenging" spirit was the commercialization of the rotary engine, which no other company in the world had ever achieved. Then, taking on the challenge of the 24 Hours of Le Mans brought this technology to the pinnacle of world motoring. In 1991, Mazda became the first rotary engine manufacturer in the world and the first Japanese manufacturer to take overall victory in this prestigious race.

The Mazda 787B had a lightweight body weighing just 830kg and a four-rotor engine with a maximum output of 700PS. The winning drivers for Mazda's 24 hours of Le Mans were Volker Weidler (German), Johnny Herbert (British) and Bertrand Gachot (French).

Direction of Mazda today

In recent years, Mazda has continued to innovate and adapt to the changing automotive landscape. The company has embraced a philosophy known as "Sustainable Zoom-Zoom," which aims to deliver both driving pleasure and environmental sustainability. Mazda has made significant advancements in fuel efficiency, with its SKYACTIV technology reducing emissions and improving performance across its vehicle lineup.

Mazda has been at the forefront of design innovation within the auto industry. The company's KODO design language, characterized by flowing lines and dynamic proportions, has garnered praise for its elegance and sophistication. Mazda's commitment to design excellence has resulted in a number of awards and accolades, solidifying its position as a design leader in the automotive industry.

Additionally, Mazda has embraced electric and hybrid technologies, recognizing the shift towards sustainable mobility. In 2020, the company unveiled its first all-electric vehicle, the Mazda MX-30, showcasing its commitment to a future of zero-emissions transportation. By combining its renowned driving dynamics with electric powertrains, Mazda aims to offer customers an exhilarating and eco-friendly driving experience.

The founding of Mazda marked the beginning of a journey filled with innovation and success. From its humble beginnings as a cork manufacturer, Mazda has evolved into a global automotive powerhouse. The company's key leaders, including Jujiro Matsuda and Kenichi Yamamoto, played instrumental roles in shaping Mazda's identity and guiding its growth.

There are several pop culture references to the Mazda car company and Mazda cars through various forms of media. Here are a few notable examples:

- "The Fast and the Furious" film franchise: Mazda cars have appeared in multiple installments of the "Fast and Furious" series. For instance, in the first movie, a Mazda RX-7 was driven by one of the main characters, Dominic Toretto (played by Vin Diesel).

- "Initial D" manga and anime series: The protagonist of this popular street racing-themed series, Takumi Fujiwara, drives a modified Mazda RX-7 FD3S. The manga and anime have gained a significant fanbase worldwide.

- "Gran Turismo" video game series: Mazda cars, including the iconic Mazda MX-5 Miata, have been featured in the long-running racing simulation game franchise. Players can race and customize various Mazda models in the game.

- *Spider-Man: Homecoming* film: In this Marvel superhero movie, the character Peter Parker (played by Tom Holland) drives a second-generation Mazda MX-5 Miata as his personal vehicle. The car appears in several scenes through the film.

- Music videos: Mazda cars have made appearances in various music videos. For example, in the music video for the song "Boulevard of Broken Dreams" by Green Day, a Mazda RX-7 can be seen parked on the street.

- "Top Gear" TV show: The popular automotive show has featured Mazda cars in several episodes. Notably, they have praised the Mazda MX-5 Miata for its handling and driving experience.

These are just a few instances where Mazda cars have been referenced or featured in pop culture. The brand's distinctive models and its association with racing and performance have made it recognizable and appealing to enthusiasts in various forms of media.

Mazda's "Zoom Zoom" marketing campaign was introduced in the early 2000s as a way to rebrand the company and create a distinctive identity in the highly competitive automotive market. The campaign aimed to emphasize the fun and exhilarating driving experience offered by Mazda vehicles.

The term "Zoom Zoom" was coined by the advertising agency Doner, which was hired by Mazda to develop a new marketing strategy. The agency wanted to create a tagline that would resonate with consumers and convey a sense of excitement and energy. "Zoom Zoom" was chosen because it was catchy, easy to remember, and could be used across various marketing channels.

The campaign was officially launched in 2000 with a series of television commercials featuring the "Zoom Zoom" tagline. These ads showcased Mazda vehicles in action, highlighting their performance, handling and stylish design. The campaign also included print advertisements, online promotions and sponsorship of various sports events, such as the American Le Mans Series.

The "Zoom Zoom" campaign was a departure from Mazda's previous marketing efforts, which focused more on practical features and affordability. By positioning itself as a brand offering more than just reliable transportation, Mazda aimed to attract a younger, more adventurous demographic.

The campaign was highly successful, helping Mazda increase its brand recognition and sales. The "Zoom Zoom" tagline became synonymous with Mazda and was often used in product brochures, dealer promotions, and even in Mazda's corporate communications.

Over the years, Mazda has continued to evolve the "Zoom Zoom" campaign, adapting it to suit different markets and product launches. The campaign has become an integral part of Mazda's brand identity, representing the company's commitment to delivering vehicles that are not only functional but also enjoyable to drive.

Today, Mazda continues to push the boundaries of automotive design and engineering, focusing on sustainability and delivering driving pleasure. With its commitment to innovation, Mazda is poised to continue to provide automotive consumers with an additional quality choice in dealer showrooms and excitement behind the wheel. Zoom Zoom!!

40

Lamborghini

Italians create modern marvel

Ferruccino Lamborghini

"The Lamborghini is a car that's designed for people who have a passion for speed and power."
—*Jay Leno*

The Lamborghini car company was founded in 1963 by Ferruccio Lamborghini in Sant'Agata Bolognese, Italy. Ferruccio was a successful businessman who had made his fortune from manufacturing tractors. He was a big fan of sports cars and owned several high-end models including a Ferrari. However, he was dissatisfied by the lack of innovation and the level of customer service provided by Ferrari, so he decided to create his own car company.

Lamborghini quickly gained a reputation for producing some of the most luxurious and powerful cars in the world. The company's first car, the 350 GT, was introduced in 1964 and was followed by several other models, including the Miura, Countach and Diablo, known for their sleek design, powerful engines and impressive speed.

Over the years, the ownership of Lamborghini has changed several times. In 1972, Ferruccio sold the company to a Swiss businessman, Georges-Henri Rossetti. The company then passed through the hands of several other owners, including Chrysler and the Malaysian company Mycom Setdco. In 1998, Lamborghini was acquired by the Volkswagen Group, which also owns other luxury car brands, including Audi, Bentley and Bugatti.

Under Volkswagen's ownership, Lamborghini has continued to produce high-performance sports cars that are popular with car enthusiasts and collectors around the world. The company has also expanded its product line to include SUVs, including the Urus, which has been a commercial success.

Lamborghini has undergone several ownership changes since its start in 1963. The company was founded by Ferruccio Lamborghini, who owned it until 1974. It was then sold to Georges-Henri Rossetti and René Leimer, who struggled to keep the company afloat and eventually sold it to Mimran brothers in 1980. The Mimran brothers were successful in turning the company around and sold it to Chrysler Corporation in 1987. In 1994, Lamborghini was sold to Indonesian company Megatech. After Megatech faced financial difficulties, Lamborghini was sold to Volkswagen Group in 1998, where it remains today as a subsidiary of Audi AG.

Famous television shows and movies that have featured Lamborghini cars include:

- *The Italian Job* (1969) - a Lamborghini Miura P400 is featured in the opening scene.

- *Cannonball Run* (1981) - a Lamborghini Countach LP400S is driven by character Jill Rivers.

- *Miami Vice* (1984-1990) - a white Lamborghini Countach, driven by Detective Sonny Crockett in several episodes.

- *The Dark Knight* (2008) - a Lamborghini Murciélago LP640, driven by Bruce Wayne.

- *Transformers: Revenge of the Fallen* (2009) - a Lamborghini Reventón, transformed into a Decepticon named "Sideways".

Some famous people who have owned Lamborghini cars:

- Kanye West
- Justin Bieber
- Jay Leno
- Chris Brown
- Floyd Mayweather, Jr.
- David Beckham
- Nicki Minaj
- Shaquille O'Neal
- Lewis Hamilton
- Cristiano Ronaldo
- Wiz Khalifa
- Deadmau5
- Arnold Schwarzenegger
- Sylvester Stallone
- Rowan Atkinson (*Mr. Bean*)
- Jerry Seinfeld
- Tim Burton
- Pierce Brosnan
- Nicolas Cage
- Tom Brady

sThe Lamborghini car company has a rich history and has become one of the most famous car brands in the world. Its success can be attributed to its founder's passion for innovation and luxury, as well as the company's commitment to producing high-performance sports cars that are both stylish and powerful. Despite changes in ownership over the years, Lamborghini has remained a top player in the luxury car market and continues to attract customers who seek the ultimate driving experience.

41

McLaren

British car with a New Zealand accent

Bruce Mclaren

*"The McLaren F1 is the finest driving machine
yet built for the public road."*
—Jeremy Clarkson, BBC Top Gear

McLaren is a British car company that was founded in 1963 by Bruce McLaren, a New Zealand racing driver who was also a designer and engineer. He started the company with the aim of designing and building racing cars. McLaren started out as a small team of engineers and mechanics who were passionate about racing and wanted to push the boundaries of what was possible.

Over the years, McLaren has become one of the most successful racing teams in history, winning numerous championships in various categories. The company has also expanded into other areas, such as building road cars and providing engineering services to other companies.

Famous television shows and movies that have featured McLaren cars include:

- *Top Gear*
- *The Grand Tour*
- *Jay Leno's Garage*
- *The Fast and the Furious* franchise
- *Cars 2*
- *The Dark Knight*
- *Need for Speed*
- *Transformers: Age of Extinction*
- *Rush*
- *Mission: Impossible: Ghost Protocol*
- *Iron Man 2*
- *Johnny English Strikes Again*
- *The Transporter Refueled*

Famous people who have owned McLaren cars over the years include:

- Jay Leno, comedian and car enthusiast, owns a McLaren F1 and has featured it on his YouTube channel.
- Lewis Hamilton, six-time Formula One World Champion, has driven for the McLaren team and owns a McLaren P1.
- Rowan Atkinson, actor known for his role as *Mr. Bean*, owns a McLaren F1 and has been involved in a few accidents with it over the years.
- Michael Fassbender, actor known for his roles in *X-Men* and *Steve Jobs,* owns a McLaren 570S.

Despite its success, ownership of McLaren has changed hands several times over the years. In 1980, Bruce McLaren died in a tragic accident while testing one of his cars. The company was then taken over by Teddy Mayer, who had been a close friend and business partner of Bruce McLaren. Mayer ran the company until 1982 when it was sold to Ron Dennis, the team manager at the time.

Under Dennis's leadership, McLaren became one of the most dominant teams in Formula One, winning multiple championships with drivers such as Ayrton Senna and Alain Prost. In 1990, Dennis formed a partnership with Mercedes-Benz, which eventually led to the German carmaker taking a 40% stake in the company.

In 2017, Dennis stepped down from his role as CEO of McLaren, and the company was sold to a consortium of investors led by businessman Mansour Ojjeh and the Bahraini sovereign wealth fund. Today, McLaren is still a major force in the world of motorsports, and the company continues to push the boundaries of what is possible in both racing and road car design.

42

Hyundai

South Koreans enter US auto market with leader

Chung Mong Koo

"Hyundai's rise to prominence in the automotive industry is a testament to their dedication to innovation and excellence."
—*Motor Trend*

Hyundai Motor Company is one of the world's largest automobile manufacturers, and it has been making cars for over five decades. The company was founded in 1967 by Chung Ju-Yung, a South Korean entrepreneur who had previously worked in the construction industry. Chung Ju-Yung was born in Tsusen, Kogen-do, Korea on November 25, 1915. Initially, the company focused on the production of small cars such as the Pony, which was launched in 1975. Despite facing numerous challenges in its early years, Hyundai has managed to become a global automotive powerhouse, with a presence in over 190 countries.

One of the key factors that has contributed to Hyundai's success is its commitment to innovation. From the outset, the company invested heavily in research and development, with the aim of producing vehicles that meet the evolving needs of consumers. Hyundai was one of the first automakers to develop an electric vehicle, with the launch of the Sonata Electric in 1991. The company has also been at the forefront of the development of hydrogen fuel-cell technology, with the launch of the Tucson Fuel Cell in 2013.

Another factor that contributed to Hyundai's success is its focus on quality. In the early years, Hyundai faced criticism for producing cars that were perceived as being of lower quality than those produced by its Japanese rivals. However, the company responded by investing heavily in quality control processes and in the training of its workforce. Today, Hyundai is known for producing cars that offer a high level of quality and reliability, and the company regularly ranks highly in customer satisfaction surveys.

Hyundai's success can also be attributed to its global outlook. Unlike many of its competitors, Hyundai has been quick to recognize the importance of expanding into new markets. In the 1980s, the company began exporting its cars to North America and Europe, and it has since expanded into other regions such as Latin America and the Middle East. Today, Hyundai has a presence in over 190 countries, and the company is constantly looking for new opportunities to expand its reach.

In recent years, Hyundai has also made significant investments in the development of autonomous driving technology. The company has partnered with a number of highly successful technology firms, including Cisco and Baidu, to develop advanced autonomous driving systems. Hyundai's goal is to become a leader in the field of autonomous driving and it has set a target of launching fully autonomous vehicles by 2025.

Despite a string of significant successes, Hyundai has faced a number of challenges over the years. One of the most monumnetal of these was the Asian financial crisis of the late 1990s. During this period, many of Hyundai's competitors were forced to scale back their operations or even went bankrupt. However, Hyundai managed to weather the storm, thanks in large part to its strong financial position and its commitment to innovation.

Another challenge that Hyundai has faced is the increasing competition in the automotive industry. Today, the new car market is more crowded than ever with new entrants including Tesla and Chinese automakers posing a threat to established players. Hyundai has responded to this challenge by continuing to invest in research and development, as well as focusing on the development of new technologies such as electric-powered and autonomous vehicles.

Looking to the future, Hyundai is well-positioned to continue its success. The company has a strong brand, a global presence and a commitment to innovation that will help it to stay ahead of the curve. However, it will need to remain vigilant in the face of increasing competition and continue to invest in new technologies in order to stay at or near the front of the pack.

The story of Hyundai Motor Company is one of perseverance, innovation and success. From its humble beginnings in the 1960s, the company has grown to become a global automotive power-house. Hyundai's success can be attributed to its commitment to innovation, quality, and global outlook. However, the company has also faced numerous challenges over the years, including the Asian financial crisis and increasing competition in the automotive indus-try. Despite these challenges, Hyundai has remained resilient, and it is well-positioned to continue its success in the years to come.

Today, Hyundai serves auto consumers with several popular models including Kona, Santa Fe, Tucson, Santa Cruz, Sonata, Ioniq 5, Ioniq 6 and Palisade.

43

Saturn

A modernized car company built inside an old car company

Saturn

"Our mission at Saturn is to challenge the status quo and push the boundaries of what a car company can be. We want to lead the industry in customer satisfaction, innovation and environmental responsibility."
—Cynthia Trudell, former president of Saturn Corporation (1995)

The creation of Saturn Motors by General Motors was a significant event in the automotive industry. It was conceived with a vision to challenge the dominant Japanese automakers, who were gaining market share in the United States during the 1980s.

General Motors aimed to create a brand that would offer innovative and fuel-efficient vehicles, focusing on customer satisfaction and a unique buying experience. The leaders within General Motors who spearheaded the creation of Saturn were Roger B. Smith, the CEO of General Motors at the time, and Alex C. Mair, who was appointed as the President of the newly formed Saturn Corporation.

Saturn was originally founded as a separate company and bought out by GM. According to Saturn, the idea for an all-new type of small car came from Alex C. Mair in summer 1982. By fall 1983, GM president F. James McDonald and chairman Roger B. Smith were on board with the idea. The first Saturn concept vehicle was released in fall 1984 and the company was officially formed in early 1985. At that time, it was a private company owned by former GM leaders but was eventually bought out by GM.

Under their guidance, a team of dedicated engineers and designers worked to bring the Saturn project to life. Saturn was envisioned as a separate division within General Motors, distinct from other GM brands, with its own manufacturing facility in Spring Hill, Tennessee. The plant was designed to be a model of efficiency, utilizing innovative manufacturing techniques such as teamwork and employee involvement.

The goal was to create a culture of quality and collaboration that would set Saturn apart from its competitors. When Saturn launched its first vehicles in 1990, it received an overwhelming response from the public. The Saturn S-Series, a compact car, was well-received for its excellent fuel efficiency, affordable price and high-quality build. The brand's commitment to customer satisfaction was evident through its "no-haggle" pricing policy and a strong emphasis on customer service.

In its early years, Saturn experienced significant success. It attracted a loyal customer base by delivering on its promises of reliable vehicles and exceptional customer service. The brand expanded its lineup to include midsize sedans, coupes, and later, SUVs. Saturn's innovative plastic body panels, dent-resistant doors and continuously evolving designs further differentiated it from other GM brands.

However, as time went on, Saturn faced challenges. The brand struggled to keep up with the rapidly changing automotive landscape and the increasing competition. Despite efforts to introduce new models, such as the Saturn Ion and Saturn Vue, the brand failed to capture the attention of consumers as it had in the past.

Further, the economic downturn of the late 2000s took a toll on General Motors as a whole, leading to a financial crisis. In a bid to restructure and stay afloat, General Motors announced the closure of Saturn in 2009. This decision was met with disappointment from Saturn enthusiasts and employees alike, as the brand had once held great promise.

The launch of Saturn created processes never heard of in automotive fiefdom. These included Saturn Homecoming events and No Haggle Pricing.

Saturn Homecomings

Saturn's "Homecoming" events were a significant part of the brand's efforts to foster a strong sense of community among its owners and enthusiasts. These events allowed Saturn owners to visit the Spring Hill, Tennessee, plant where their cars were manufactured and connect with the people behind the brand. The first "Homecoming" in 1994 attracted 44,000 people. Owner enthusiasm went off the charts, as was demonstrated when nearly 100,000 owners attended two "homecoming" celebrations in 1994 and 1999.

Here is an overview of what the Saturn Homecoming typically entailed:

1. Purpose: Saturn Homecoming events were organized to celebrate the Saturn community, connect owners with each

other, and provide an opportunity for Saturn enthusiasts to see how their cars were made.

2. Location: The events were held at the Spring Hill Manufacturing facility, which was Saturn's primary manufacturing plant. Spring Hill was chosen as it represented the heart of the Saturn brand.

3. Plant Tours: One of the highlights of the Homecoming events was the opportunity for attendees to take guided tours of the Spring Hill plant. These tours allowed visitors to witness the manufacturing process firsthand and gain a deeper appreciation for the brand's commitment to quality.

4. Meet the Team: Saturn Homecomings often featured meet-and-greet sessions with Saturn employees, including assembly line workers, engineers, and executives. This interaction gave attendees a chance to ask questions and learn more about the people responsible for building their cars.

5. Car Show: Owners often had the opportunity to showcase their own Saturn vehicles at car shows during the Homecoming events. This allowed enthusiasts to share their pride and passion for their cars with fellow Saturn owners.

6. Workshops and Seminars: Saturn sometimes organized workshops and seminars during these events. These sessions could cover topics such as car maintenance, customization, and insights into Saturn's unique manufacturing processes.

7. Entertainment: Saturn Homecomings were not just about cars; they often featured entertainment, live music, and family-friendly activities to create a festive atmosphere.

8. Merchandise and Memorabilia: Attendees could typically purchase Saturn-branded merchandise and collectibles to commemorate their visit.

9. Community Building: The Homecoming events helped build a tight-knit community of Saturn owners and enthusiasts who shared a passion for the brand. It created a sense of belonging and loyalty among Saturn owners.

Saturn's Homecoming events were seen as a testament to the brand's commitment to customer engagement and community building. Unfortunately, with the discontinuation of the Saturn brand in 2010, these events became part of the brand's history. However, they remain fond memories for Saturn owners and enthusiasts who were fortunate enough to attend and experience the unique Saturn spirit firsthand.

No Haggle Pricing

Saturn's no-haggle price model immediately gained popularity with buyers both in the U.S. and Canadian markets. Saturn celebrated the release of its 500,000th car, which the company named Carla, in 1993, and its millionth car in 1996. Saturn had one of the highest customer loyalty rates in the industry. Their No Haggle Pricing was wildly popular with female buyers.

GM's first electric car

GM's first-ever electric car was the Saturn GM EV1. Like many others, inside and out of the industry, I was intrigued by Saturn. I wondered about its processes, innovation and long term, practical strategies toward future industry change and success. During an unrelated planned road trip between a conference in New Orleans and a recreational event in Southern Indiana, I planned my driving route in order to make a swing through Spring Hill, Tennessee, just a few miles and minutes south of Nashville.

I found that that it was enlightening to see that within an old, somewhat stoic corporation like General Motors, it is possible to innovate and think differently. For that reason, and others, I was disappointed to hear that General Motors, under pressure from the Obama Administration's famous Auto Task Force, added Saturn to the short list of brands to be terminated during the bankruptcy filing in May of 2009.

As the industry evolved, Saturn struggled to adapt, eventually leading to its closure in 2009 amidst financial difficulties faced by General Motors. Despite its ultimate fate, Saturn Motors will always be remembered as an ambitious venture that aimed to revolutionize the automotive industry.

44

Tesla

From its 21st Century beginnings

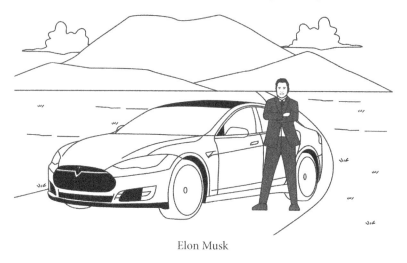

Elon Musk

"I'm highly confident that we'll be able to create a car that is superior to the best gasoline cars in every way. That's our goal."
—*Elon Musk*

Tesla, an American-based automaker which some call a tech company, has become synonymous with innovation and disruption in the automobile industry. Founded in 2003, Tesla may be the first start-up volume automaker to become successful in the past 100 years; the company has relatively quickly become the most valuable automaker in the world, based strictly on stock value. Today, Tesla is known for producing higher-end electric vehicles, but the company's beginnings were much humbler.

Over the past 20 or so years, I have been one of Tesla's most vociferous critics. I remain a critic today, though now focused mostly on the very poor service accessibility, the too-fast rush to push autonomous features (Autopilot and Full Self Drive – FSD), as well as Tesla's frequent push to break the rules, not just bend them. Avoidance of and ignorance to 'playing by the rules' has been a historic trait of Elon Musk and thus Tesla, it seems. Yet, I give the company credit for doing almost the impossible—successfully entering the hard-to-break-into new car industry in modern times. Here's my (fair) Tesla take.

Tesla was founded by a group of engineers, led by Martin Eberhard and Marc Tarpenning. Eberhard was born in California in 1960 and grew up in the San Francisco Bay Area. Eberhard attended the University of Illinois, where he earned a degree in computer engineering. After graduation, Eberhard worked for a variety of tech companies, including Wyse Technology and Network Computing Devices.

Tarpenning was born in Sacramento in 1964, though he grew up in Southern California. Tarpenning attended the University of California, Berkeley, where he earned a degree in computer science. After graduation, Tarpenning worked at a variety of tech companies, including HP. It was during his time at HP that Tarpenning first met Eberhard.

Late in the 1990s, Eberhard and Tarpenning began discussing the big, bold idea of starting their own automotive company. They were both interested in electric vehicles and saw an opportunity to create a new kind of car company that would focus on vehicles driven with battery electric power. At the time, the automobile industry was dominated by large, established companies such as General Motors, Ford and Toyota. Eberhard and Tarpenning believed there was room for a new automotive player in the market, one that would focus on technology and innovation.

In 2003, Eberhard and Tarpenning founded Tesla Motors. The company's name was inspired by the famous inventor and electrical engineer, Nikola Tesla. Tesla was a pioneer in the field of electricity and is known for his contributions to the development of the alternating current (AC) electrical system.

At the time of its founding, Tesla was a very small company with just a handful of employees. Eberhard and Tarpenning were co-founders and Co-CEOs of the company. They were joined by a team of engineers who shared their passion for electric vehicles.

The early days of Tesla were marked by a lot of hard work and uncertainty. The company was focused on developing its first vehicle, the Tesla Roadster, which was a high-performance electric sports car. The Roadster was the first production car to use lithium-ion battery cells and came with a driving range of over 200 miles on a single charge. The car was a hit with early adopters and helped to establish Tesla as a serious player in the electric vehicle market.

Despite the limited production yet success of the Roadster, Tesla faced a number of challenges in its early years. The company struggled to raise capital and had to rely on a series of private funding rounds to keep the business afloat. Eberhard and Tarpenning eventually stepped down from their roles as co-CEOs and were replaced by Elon Musk, who had invested in the company and became its largest shareholder.

Under Musk's guidance, Tesla has grown into a global powerhouse within the automobile industry. The company expanded its product line to include a range of electric vehicles, starting with the Model S, Model X, Model 3 and Model Y. Musk liked to use the vehicle model identifying letters together to spell out "S3XY." In addition to producing cars, following the acquisition of Solar City, Tesla developed a range of energy products, including solar panels and battery walls for homes and businesses.

Though the story of Tesla Motors began in 2003, before the first Tesla Roadster was produced, the company had to overcome a number of challenges, including building manufacturing facilities and developing the technology that would power the cars.

In the early days of Tesla, from the conception of the company to the development of the first car in the lineup, the Roadster, the company faced many challenges. How the challenges were overcome as well as the impact that has led to much mystic, even cult-like fan-base following for both Elon Musk as chairman and Tesla as an automaker or technology company, depending on what it really is.

Building the manufacturing facilities

One of the biggest challenges Tesla faced in its early days was building the manufacturing facilities needed to produce electric vehicles. The company had to find a location large enough to house a production line with access to the resources needed to build cars.

After considering a number of different locations, Tesla settled on a site in Fremont, California, that had been used as a jointly owned factory by General Motors and Toyota. The site was already equipped with many of the resources that Tesla needed, including a paint shop and an assembly line, which helped to reduce the initial cost of building all the capabilities of its facilities.

Even with the existing infrastructure, Tesla had to invest a significant amount of money in upgrading the facilities to meet its needs. New equipment was installed and the production line was reconfigured to accommodate Tesla's unique manufacturing processes.

Developing the technology

Another significant challenge was developing the technology needed to power the cars. The company had to develop new battery technology that would be efficient, reliable and long-lasting, as well as an electric motor that would provide the performance that customers expected from a sports car.

To accomplish this, Tesla partnered with a number of companies, including Panasonic, to develop new battery technology. It also hired a team of engineers to design and build a new electric motor that would be powerful enough to propel its cars while also being compact enough to fit into a small sports car. Following the launch of the Roadster, came the following models: Model S, Model X, Model 3 and Model Y. Each helped Tesla build greater volume sales and expand nearly worldwide.

Beyond the first factory in Fremont, California, Tesla built a Gigifactory for battery production near Reno, Nevada. Later, a Tesla car factory was added in China. Another US factory was added near Austin, Texas. And a new factory was built and brought online near Berlin in Germany.

Tesla also worked to release an all-electric Semi, which started slowly mass producing in mid-2023. And after four years of promises, Tesla rolled out its CyberTruck, an unusual looking, all stainless-steel pickup that is large in structure yet small in cargo-carrying bed-size.

What sets Tesla apart from more traditional automakers?

The irony of Tesla as a relatively new automaker primarily links to its exorbitant company value based on cumulative stock price. For most of the past five years, Tesla's inflated corporate value has exceeded that of the next 10 largest automakers combined. Some think that is just insane. Count me in that group of critics, what some (okay, many) call Tesla bears.

Tesla has effectively set itself apart from traditional automakers by building out the largest network of public electric charging stations, which the company calls Tesla Superchargers. Tesla has built 12,000 Supercharger stations nationwide, an amount far greater than any other automaker. In fact, it is believed that the Tesla Supercharger network is larger than all other manufacturer's fast chargers collectively.

The success and vision of Tesla's supercharging EV charging network has revolutionized the electric vehicle industry and has played a crucial role in making electric vehicles more accessible and convenient for consumers. The following paragraphs explore the reasons behind the success of Tesla's supercharging network and its vision for the future.

Tesla's supercharging network was introduced in 2012 as a way to address one of the main concerns of electric vehicle owners: range anxiety. Range anxiety refers to the fear of running out of battery power while on the road and not having access to a charging station. Tesla recognized this concern and set out to create a charging network that would alleviate these fears and make long-distance travel in an electric vehicle a viable option.

One of the key factors behind the success of Tesla's supercharging network is its extensive coverage. Tesla has strategically placed supercharger stations along major highways and in urban areas, ensuring that electric vehicle owners have access to charging infrastructure wherever they go. This extensive coverage has made it possible for Tesla owners to travel long distances without worrying about running out of battery power.

Furthermore, Tesla's supercharger network offers fast charging speeds, allowing electric vehicle owners to charge their vehicles quickly and efficiently. The latest generation of superchargers can provide up to 250 kW of charging power, enabling Tesla vehicles to add up to 200 miles of range in just 15 minutes. This fast-charging capability has significantly reduced the time required for charging, making electric vehicles more convenient for everyday use.

Another reason for the success of Tesla's supercharging network is the company's commitment to continuous improvement and innovation. Tesla has been investing heavily in research and development to improve the charging experience for its customers. For example, the company has introduced features like V3 Supercharging, which further increases charging speeds and reduces charging times. Tesla is also working on expanding the network's capacity to accommodate the growing number of electric vehicles on the road.

In addition to its current success, Tesla has a clear vision for the future of its supercharging network. The company aims to make charging even more convenient and accessible by further expanding the network and increasing charging speeds. Tesla plans to double the number of supercharger stations globally by the end of 2024, ensuring that electric vehicle owners have even more options for charging their vehicles. The company is also working on deploying its superchargers in urban areas, allowing Tesla owners to charge their vehicles while they shop or dine.

Tesla's supercharging network has not only been successful in terms of its impact on the electric vehicle industry but also in terms of its contribution to sustainability. By providing a reliable and convenient charging infrastructure, Tesla has encouraged more people to switch to electric vehicles, reducing the dependence on fossil fuels and decreasing carbon emissions. The success of Tesla's supercharging network has paved the way for other automakers to invest in their own charging infrastructure, further accelerating the adoption of electric vehicles.

The success of Tesla's supercharging EV charging network can be attributed to its extensive coverage, fast charging speeds, continuous improvement and clear vision for the future. Tesla has addressed the concerns of range anxiety and made electric vehicles more accessible and convenient for consumers. With its commitment to expanding the network and developing new technologies, Tesla is leading the way towards a sustainable and electric future.

Challenges for Tesla customers with build quality and service scheduling

While Tesla cars generally receive positive reviews, like any other brand, they are not without their share of complaints. Some of the most commonly reported issues by Tesla owners include:

1. **Build Quality:** Some consumers have reported issues related to fit and finish, such as panel gaps or misalignments. However, it's worth noting that the severity of these issues can vary from car to car, and Tesla has been working to improve its manufacturing process over time.

2. **Service and Repairs:** Some customers have expressed dissatisfaction with Tesla's service and repair process, including long wait times, difficulty in finding authorized repair centers, and delays in receiving replacement parts.

3. **Software and Firmware Updates:** While Tesla's over-the-air software updates are often praised for introducing new features and improvements, some customers have reported occasional bugs, glitches or issues after updating their car's software.

It's important to note that the experiences and opinions of Tesla owners can vary widely. While some may have encountered issues, others have had positive experiences with their vehicles and praise Tesla's build quality and finish. Every automotive consumer should always thoroughly research and consider multiple sources before making any purchasing decisions.

Autopilot and full self-driving technologies

The advent of autonomous driving technology has brought about a significant transformation in the automotive industry. Tesla, a leading electric vehicle manufacturer, has made remarkable strides in this field with its Autopilot and Full Self Drive (FSD) features. While these technologies offer numerous benefits, they are not without their drawbacks. The following paragraphs explore both the positives and shortcomings of Tesla's Autopilot and FSD, highlighting their potential to revolutionize autonomous driving.

Positives of Tesla Autopilot and FSD

- **Enhanced Safety and Reduced Accidents**
  One of the primary advantages of Tesla's Autopilot and FSD is their potential to significantly enhance road safety. These technologies utilize an array of sensors, cameras and advanced algorithms to monitor the vehicle's surroundings and make real-time decisions. By constantly analyzing data, Tesla vehicles equipped with Autopilot and FSD can detect and respond to potential hazards faster than human drivers, reducing the likelihood of accidents caused by human error.

- **Advanced Driver Assistance Features**
  Tesla's Autopilot and FSD provide advanced driver assistance features that make driving more convenient and enjoyable. Autopilot allows for adaptive cruise control, automatic lane changes, and self-parking, making long drives less tiring and reducing driver fatigue. Furthermore, FSD aims to offer a fully autonomous driving experience, enabling users to summon their vehicles remotely and navigate complex traffic situations effortlessly.

- **Continuous Software Updates and Improvements**
  Tesla's Autopilot and FSD systems benefit from over-the-air software updates, which enable the company to continuously improve and enhance the capabilities of these technologies. This means that even vehicles already on the

road can receive new features, performance improvements, and safety enhancements, keeping the technology up-to-date and ensuring a future-proof driving experience.

- **Data Collection and Machine Learning**
  Tesla's fleet of vehicles equipped with Autopilot and FSD serves as a massive data collection network. This data is anonymized and used to train machine learning algorithms, allowing Tesla to refine and improve the performance and safety of their autonomous driving technology. The more data the system accumulates, the better its decision-making capabilities become, creating a positive feedback loop that benefits all Tesla drivers.

Shortcomings of Tesla Autopilot and FSD

- **Limitations in Complex Driving Conditions**
  Despite its remarkable capabilities, Tesla's Autopilot and FSD still face limitations, particularly in complex driving situations. These technologies heavily rely on pre-mapped data and struggle to adapt to rapidly changing environments, such as construction zones, poorly marked roads, or adverse weather conditions. Consequently, drivers must remain vigilant and be ready to take control of the vehicle at any moment, especially in situations where the system may falter.

- **Overdependence on Technology and Complacency**
  The convenience and assistance provided by Tesla's Autopilot and FSD can lead to overdependence on the technology, causing drivers to become complacent and less attentive. This complacency may result in delayed reaction times when the system requires human intervention, potentially leading to accidents. It is crucial for drivers to remain

engaged and fully aware of their surroundings, treating Autopilot and FSD as driver assistance tools, rather than fully autonomous systems.

- **Regulatory and Legal Challenges**
  The widespread adoption of autonomous driving technology poses significant regulatory and legal challenges. As Tesla pushes the boundaries of autonomous driving, it encounters various legal and regulatory hurdles, including questions surrounding liability in case of accidents, compliance with traffic laws, and the need for clear regulations to govern the use of these technologies. Resolving these challenges will require collaboration between automakers, regulators, and legal institutions to ensure the safe and responsible integration of autonomous driving technology.

- **Ethical Dilemmas and Decision Making**
  Autonomous driving technology raises complex ethical dilemmas when it comes to decision-making in critical situations. For instance, in cases where an accident is imminent, the system may need to make split-second decisions that prioritize the safety of the vehicle's occupants or other road users. Determining the appropriate decision-making algorithms and ethical guidelines for autonomous vehicles is a significant challenge that requires careful consideration and public discourse.

Tesla's Autopilot and Full Self Drive (FSD) technologies have the potential to revolutionize the automotive industry and pave the way for a future with safer and more efficient transportation.

Tesla as a wrap

Without doubt, even Tesla critics, such as me, must give the company its due. Part of that, and maybe most of it, is related to Elon Musk's tenacity and bold business acumen. I still think the brash nature of the way Tesla does business will change over time. It has to. What goes up, must come down. Though I am no longer one that predicts Tesla will fail, at least not in the foreseeable future. Though like any established automaker, Tesla will see rough waters ahead. Either based on quality of product, lack of ease of service along with service wait times, changes in public sentiment or other, Tesla will hit choppy waters. As the saying goes, its not how many times you go down that matters. It's how many times you can get back up.

45

Rivian

RJ Scaringe and Rivian: In the shadow of Tesla, pioneering the electric vehicle revolution

RJ Scaringe

"We have an incredible opportunity to create a new electric vehicle company from the ground up and change the way people think about transportation."
—*RJ Scaringe, Founder and CEO of Rivian*

In recent years, the auto industry has driven a rapid shift towards sustainable transportation solutions and electric vehicles (EVs) have emerged as an important component within this revolution. Among the visionary leaders propelling this change is RJ Scaringe, the founder and current CEO of Rivian, an emerging American electric vehicle manufacturer. Here we explore the journey of RJ Scaringe and Rivian, examining their innovative approach to EV production and the remarkable success trajectory Rivian has achieved to date.

RJ Scaringe, the visionary behind Rivian is an automotive enthusiast and entrepreneur. He founded Rivian in 2009 with a concept to create sustainable mobility solutions that redefine the automotive industry. With a background in mechanical engineering and a Ph.D. in automotive engineering from the Massachusetts Institute of technology (MIT), Scaringe encompassed the technical expertise needed to revolutionize the electric vehicle market.

The concept for Rivian was born from what was Mainstream Motors, a company that was focused on developing a fuel-efficient gas-powered vehicle. However, Scaringe quickly recognized the potential of electric vehicles and shifted the company's focus towards EV development. Scaringe's strategic decision helped lay the foundation for Rivian's long-term success.

Rivian's initial product lineup includes the R1T, an all-electric pickup truck, and the R1S, a seven-passenger electric SUV. These vehicles incorporate advanced battery technology, impressive range capabilities and cutting-edge features, making them highly attractive and competitive in the fast-growing electric vehicle market. Rivian established a commitment to sustainability that extends beyond its vehicles as the company has also invested in developing a network of charging stations and sustainable manufacturing processes.

To bring his vision to life, Scaringe sought significant investments to fund Rivian's ambitious projects. Notably, in 2019, the company secured a $700 million investment led by Amazon, followed by a $500 million investment by Ford Motor Company. These strategic partnerships not only provided financial support but also offered access to invaluable resources and expertise which helped trigger Rivian's growth.

Since its original launch, Rivian experienced a remarkable success trajectory, solidifying its position as a frontrunner in the electric vehicle industry. The company's innovative approach, commitment to sustainability and commendable focus on customer-centric design have garnered widespread acclaim.

Rivian's vehicles combine sleek design, cutting-edge technology, and exceptional performance. The R1T and R1S have received rave reviews for an impressive driving range, off-road capabilities and other advanced technology features. Rivian's focus on creating versatile vehicles that cater to both adventure enthusiasts and urban dwellers has resonated with prospective consumers worldwide.

Rivian's success is not only evident in its product offerings but also in the market demand for its vehicles. The R1T and R1S garnered sizable numbers of pre-orders with early adopters eagerly awaiting their delivery. The company's ability to compete with established automotive giants, such as Tesla, highlights its strong market presence and consumer confidence in the brand.

Rivian's commitment to sustainability extends beyond producing electric vehicles. The company has implemented eco-friendly manufacturing processes and is actively involved in environmental conservation efforts. Additionally, Rivian's focus on building a network of charging stations across the United States promotes EV adoption and reduces range anxiety further contributing to the company's early success.

Challenges for Rivian now and into the future

As did Tesla before it, Rivian chose to distribute to consumers directly though its factory-owned, exclusive closed manufacturer network versus building out an independent franchised dealer distribution network. There are advantages and disadvantages to each distribution network model. For the automaker, choosing to distribute through a factory-owned and controlled system increases their cost of operation. To cover the country effectively (not to mention the world) that means a cost of $20 million to $40 million for each service center (or in the franchised model, call those dealership locations).

When that cost is multiplied by 300—based on typical luxury brand franchised dealer locations—that can add $9 billion in operating investment. It also adds a monumental challenge in staffing, managing, operating and oversight to be able to successfully lead 300+ locations in ways that create sought-after and commendable customer service.

Further, with franchised dealers, both inter-brand and intra-brand competition for better pricing and negotiating power by the consumer to get the price in a place that works. With the factory-distribution system, there is only the MSRP vehicle purchase price available. Since Tesla had 13 price cuts in the first half of 2022 alone, factories have proven themselves to be price knowledgeable. Yet it is the manufacturer who makes the decision and not necessarily the consumer.

Future outlook and impact

As Rivian continues to innovate and expand its product lineup, the company's future outlook appears promising. With plans to launch additional models and expand its global presence, Rivian is appropriately situated to make a significant impact on the automotive industry. The success of Rivian could serve as an inspiration for other entrepreneurs and automakers to embrace sustainable mobility solutions and invest in the future of electric transportation.

RJ Scaringe's visionary leadership and Rivian's commitment to sustainable mobility have positioned the company as a trailblazer in the electric vehicle industry. Through innovative design, cutting-edge technology and strategic partnerships, Rivian has achieved remarkable success to date. As the demand for electric vehicles continues to rise, Rivian's consistent commitment to environmental sustainability and customer-centric design will undoubtedly ensure its continued success, helping drive the automotive industry towards a greener and more sustainable future.

46

Lucid

Doing its part to revolutionize the EV industry

Peter Rawlinson

"Lucid is dedicated to pushing the boundaries of electric vehicle technology and delivering an unrivaled driving experience."
—*Peter Rawlinson, Founder, CEO and CTO of Lucid Motors*

Lucid Motors, an electric vehicle (EV) startup, has gained significant attention and recognition for its innovative approach towards sustainable transportation. This chapter provides an in-depth analysis of Lucid, covering various aspects including its founders, background, early and current investors, home office, factory locations, production timeline, dealership network, future models, pricing, profitability, and prospects for success.

Lucid Motors was founded in 2007 by Bernard Tse, Sam Weng, and Peter Rawlinson. Bernard Tse, a former Tesla executive, played a crucial role in the early development of Tesla's Roadster. Sam Weng is an entrepreneur with a background in technology start-ups who brought invaluable experience in managing high-growth companies.

Peter Rawlinson, the company's Chief Technology Officer, is an automotive industry veteran with extensive experience at Jaguar, Lotus, and Tesla.

Lucid received early capital through its initial investors included Mitsui & Co., Venrock, and Tsing Capital. These investments allowed the company to advance its research and development efforts and move closer to its goal of producing high-performance electric vehicles. Lucid's current investors include the Saudi Arabian Public Investment Fund. Lucid has become attractive to other investors as well.

In 2020, the company entered into a merger agreement with Churchill Capital Corp IV, a special purpose acquisition company (SPAC) led by Michael Klein. This merger transaction provided Lucid with approximately $4.4 billion in funding, enabling the company to accelerate its growth plans and expand its operations.

Lucid Motors is headquartered in Newark, California, and the company's home office serves as the central hub for its operations, research and development, and administrative functions. The first Lucid Motors factory, known as Lucid AMP-1 (Advanced Manufacturing Plant), is located in Casa Grande, Arizona. This state-of-the-art facility spans over one million square feet and is dedicated to the production of Lucid's flagship model, the Lucid Air.

In addition to the Casa Grande factory, Lucid has announced plans for a second manufacturing facility, Lucid AMP-2, set to be constructed in Saudi Arabia. The Saudi Arabian factory will cater to the growing demand for electric vehicles in the Middle East and other international markets.

The first Lucid cars to come off the assembly line hit that significant milestone on September 28, 2021, when the first Lucid Air luxury electric sedans rolled out of the Casa Grande factory. This achievement marked the beginning of Lucid's commercial production and symbolized the company's commitment to delivering high-quality electric vehicles to customers worldwide.

During visits to Phoenix in January of 2020 and again in January 2021, I was able to make quick road trips to Casa Grande to check on the progress of construction of the Lucid factory there. In January 2020, the construction had just started though you could tell where the factory would be built. At the time the building was already framed in and construction was under way. One year later, January 2021, I was back and able to see the completed Lucid factory. During the 2021 visit, I also drove a few miles out of the way to the east in order to go by Coolidge, Arizona, site of Nikola Truck's first battery electric truck manufacturing and assembly plant.

In order to ensure a seamless customer experience, Lucid Motors adopted a direct-to-consumer sales model. As of now, the company operates about 40 factory-owned dealerships strategically located in key markets. The number of dealerships is expected to grow as Lucid expands its product line and market presence. Dealerships are concentrated in areas targeted for high demand for EVs with several each in California, Florida, New York, and Texas.

New Lucid models planned, in addition to the Lucid Air, include an electric SUV and a range of electric vehicles targeting different market segments. Having started production at the high end of the auto market, pricewise, Lucid plans to work down market toward more affordable market entry models. Yet, Lucid also aims to compete aggressively in the ultra-luxury electric vehicle segment, targeting customers who prioritize style, performance and sport.

47

Gone Trucking

How American truck makers thrive under global ownership

Modern semi-trucks serve as personal mobility for the drivers and, ever increasingly, the driver's spouse or family. The sleeper units operate like RVs and the utility of semis enables the transportation network to ensure cars get to consumers along with everything else

Imagine the United States as a body, with the highway system as its veins and arteries. The blood is made up of medium- and heavy-duty trucks, delivering the goods we need while carrying our waste away.

The precursors of trucks were horse-drawn wagons, stagecoaches, and farm equipment. But trucks actually evolved from automobiles, specifically, the Winton automobile.

The Winton Motor Carriage Company hand-built automobiles beginning in 1897. Alexander Winton raced his 10-horsepower car around a track in Cleveland at the awe-inspiring speed of 33.6 miles per hour and later drove from Cleveland to New York City. He drew buyers, but how would the motor carriages be delivered? The solution was to mount a trailer on a Winton model truck, creating an auto hauler.

If automobiles could be hauled from producers to sellers, so could other goods. By 1914 more than 100,000 trucks were rolling. A handful of states enacted weight requirements, even though trucks were still constrained by speed limits and solid tires. Truck drivers formed the first union in 1901; the Teamsters Union followed in 1903.

Charles Fruehauf was hired to build a boat trailer in 1914. His client was thrilled with the result and asked Fruehauf to make trailers for his lumber yard. The idea caught on with other industries. Long-distance trucking was made viable with inflatable tires, developed during World War I. More than a million trucks plied the roads by the early 1920s. The introduction of much more efficient diesel engines came in the 1930s, followed by power-assisted steering and brakes.

Trucks, it turned out, were useful for all sorts of tasks. Their growing presence led to several developments, such as refrigerated trucks in 1938, emergence of truck stops in the '40s, the invention of the CB radio in 1945, and the construction of the Interstate Highway System beginning in 1956.

What are they?

Trucks generally fall into essentially eight (8) classes.

- Classes 1-3 include pickups, vans and SUVs – mostly non-commercial vehicles, although some Class 3 pickups qualify as medium duty.
- Class 4 (14,000-16,000 pounds) – e.g., delivery trucks.
- Class 5 (16,000-19,500 pounds) – e.g., bucket trucks, farming equipment.

- Class 6 (19,500-26,000) pounds – e.g., beverage trucks.
- Class 7 (26,000-33,000 pounds) – e.g., garbage trucks
- Class 8 trucks (33,000 pounds +) – e.g., tractor-trailer/ semis; also cement and dump trucks.
- Classes 4-8 generally require special training and driver's licenses – Commercial Drivers Licenses (CDL) are required starting at 26,000 pounds gross vehicle weight (GVW)

Truck fast facts

- The American Trucking Association (ATA) reported that 11.92 billion tons of freight were transported by trucks in 2019, representing 80.4% of the nation's freight.
- Medium and heavy-duty trucks accounted for 26% of all U.S. trucks in 2018.
- There were 3.91 million Class 8 trucks operating in 2019, including both tractor-trailer/ semi-trailer trucks and straight (single chassis) trucks.
- America's heavy-duty truck drivers traveled more than 432 billion miles in 2018, using 54 billion gallons of fuel, according to the U.S. Dept. of Transportation.
- There were 3.6 million truck drivers employed in 2019 and 795 million people in jobs related to trucking (excluding those self-employed) – about six percent of all full-time jobs in America.
- While the trucking industry is made up of 928,647 private motor carriers, the overwhelming majority of them (91.3 percent) are small businesses with five or fewer trucks.
- U.S. trucking industry's 2017 revenues exceeded the Gross Domestic Product (GDP) of more than 150 nations.
- Commercial trucks paid more than $41 billion in state and federal highway user taxes in 2015, according to the ATA.

While there are scores of manufacturers, just a handful dominate heavy- and medium-duty truck manufacturing. Freightliner (Daimler) leads the pack with 29%, followed by Ford (19%), International (13%), Kenworth (11%), Peterbilt (7%) and Volvo (5%). Others continue to make notable vehicles, though, including Mack, GMC/Chevrolet, Ram, Isuzu and Hino. Legendary status belongs to at least one no longer built: the REO.

REO

Ransom Eli Olds was successful building cars with his original company, Olds Motor Works, which became part of General Motors in 1908. REO Motor Truck Co. was formed as a subsidiary of his later company, REO Motor Car Company, in 1910. It mounted truck bodies on modified auto chassis to create three-quarter-ton trucks. Heavier-duty trucks followed beginning in World War I.

The famous REO Speed Wagon was introduced in 1915 and manufactured until 1949. It is widely considered the precursor to the pickup truck. In the 1920s, the Speed Wagon pioneered electric starters and lights, shaft-driven axles and steel-mounted pneumatic tires. The product line grew to include trucks that could carry up to two tons.

Experiencing financial problems, REO declared bankruptcy in 1938. It reorganized and supplied military vehicles during World War II. After the war, REO manufactured heavier-duty trucks, until it was sold in 1957 to White Trucks Motor Company, forming Diamond REO. The company and its iconic name changed hands again to a Class 8 truck maker with plans to continue building REO's C-116 Giant but it was not successful and REO ceased to exist in 1995.

International: from farm to trucks

Cyrus McCormick developed the horse-drawn reaper in the 1830s and co-founded the McCormick Harvesting Machine Company in

Chicago in 1847. After a series of mergers it became International Harvester in 1902.

While continuing to manufacture tractors, IH moved into vans and trucks. Beginning with the Auto Wagon, the company built light trucks from 1907 to 1975. It was also an early producer of medium- and heavy-duty trucks. The cab-over-engine (COE) heavy-duty CO-4000 – the first entirely designed and built by IH – was introduced in 1965. Through the 1980s, IH trucks were characterized by a variety of names ending in "star." Beginning with the Loadstar, there was the Class 8 Transtar, Paystar, Cargostar and Fleetstar.

IH's finances were precarious throughout the 1960s and '70s and it was substantially affected by labor problems in the '80s. It sold off pieces of its business but kept the truck and engine divisions, changing its name to Navistar International in 1986. Navistar's subsidiary, International Truck and Engine Corporation, continues to build and sell trucks and engines badged with the International name. The Traton Group, based in Germany, bought Navistar in 2021. The Traton Group owns all of Navistar stock.

White's rise and fall

White Trucks grew out of the White Motor Car Company, which in turn grew out of a business that manufactured everything from roller skates to sewing machines. After World War I, the company stopped making cars and focused on a range of trucks. Following World War II, it concentrated on heavy-duty trucks and acquired a string of competitors, including REO. It distributed Freightliner Trucks from 1951-77. Sales started dropping in the '60s and White began to have financial problems. When White filed for bankruptcy in 1980; Volvo acquired its American assets.

Mack: the industry's bulldog

Carriage-makers Jack and Gus Mack founded Mack Trucks in 1900 in Brooklyn, NY. Jack was reported to be inspired by riding in a neighbor's automobile.

An early bus builder, Mack's 40-hp, 20-passenger sightseeing bus operated in Brooklyn for several years before being converted into a truck. Mack produced the first motorized hook-and-ladder fire truck in 1910.

Mack adopted the COE (cab-over-engine) configuration in 1905. It can also claim credit for features that prevented stripped gears and allowed drivers to shift gears without going through intermediate gears. It introduced air and oil filters and rubber mounts in 1918, power brakes using a vacuum-booster system in 1927 and four-wheel brakes in 1936. Mack pioneered the all-fiberglass, metal cage-reinforced cab in the MH Ultra-Liner in 1982.

Mack's bulldog trademark originated during WWII, when the truck's blunt-nosed hood and durability reminded British soldiers of their bulldog symbol. The war effort required 35,000 Mack Trucks.

Mack became a wholly-owned Renault subsidiary in the early '80s and subsequently was bought by AB Volvo.

Peterbilt and Kenworth: a family affair

Peterbilt and Kenworth compete as separate brands though both are owned by the Pacific Car and Foundry (PACCAR) conglomerate.

William Pigott Sr. founded Seattle Car Manufacturing in 1905, soon merging with Twohy Brothers to become Pacific Car and Foundry Co., which stayed in the Pigott family for many years. The company acquired Kenworth Motor Truck Co. in 1945 and rebranded as PACCAR in 1972.

Peterbilt started as a solution to T.A. Peterman's problem in the 1930s: How to move logs from forest to mill? Floating them downriver or hauling them by steam or horse would be too slow.

He first used Army surplus trucks. Then he bought a failing motor company in 1938, establishing Peterbilt to build a truck chassis. His relentless pursuit of improvement and innovation continued until he died in 1944. Employees bought and continued to grow Peterbilt until 1958. Pacific Car and Foundry purchased Peterbilt when Peterman's widow sold the property where the factory was located.

Peterbilt's innovations include using aluminum to reduce weight (1945); the all-aluminum tilt hood (1965); the first Smartway Designated Alternative Fuel Vehicle and the first standard Air Disc Brakes (2011). It announced plans in 2018 to produce all-electric semis. Peterbilt hit the million-truck mark in 2018 with the Peterbilt Model 567 Heritage.

Seattle-based Kenworth is named for both of its founders: Edgard Worthington and Captain Frederick Kent. They acquired Gerlinger Motor Car Company in 1917, renaming it Kenworth Motor Truck Company in 1923. Kenworth expanded to building buses in 1927 and in 1932 built its first fire truck. It became the first American truck builder to make diesel engines standard in 1933; diesel was a third the cost of gasoline. It also introduced the first factory-made sleeper cab in 1933.

Kenworth supplied American forces with almost 2,000 vehicles during World War II and has continued to contract with the government since. It also supplied components for Boeing's B-17 and B-29 bombers.

Pacific Car & Foundry acquired Kenworth in 1945 when employees couldn't get financing to buy their company. Kenworth grew from a regional to an international truck manufacturer by 1950, selling trucks for Middle East oil production. In 1959 it expanded manufacturing to Mexico and later to Australia. By its 50th anniversary in 1973, annual sales had grown to 10,000 units. Kenworth continues to manufacture trucks as a PACCAR company.

The Swedish invasion

Volvo has continually sought to improve safety and fuel efficiency in both its cars and trucks.

It has manufactured trucks in Europe since 1928, moving into the U.S. market in 1971. Volvo launched the F613 medium-duty truck in 1976 and in 1981 bought part of the White Motor Corp. A joint venture between Volvo and General Motors—Volvo GM Heavy Truck Corp.—was formed in 1988. In 1997, Volvo bought GM's truck division to become Volvo Trucks North America. With its Renault and Mack Trucks, Volvo is the second-largest truck builder in the world. The parent company in Sweden, AB Volvo, sold its passenger car division to Ford Motor Company in 1999. Ford later sold the Volvo passenger car company to China-based Geeley. Since 2001 the truck company has used the Volvo brand name.

Freightliner: American with a German accent

Freightliner launched itself into the truck market in 1942 with its all-aluminum cab. Truck-building halted during World War II while Freightliner manufactured military equipment. It introduced the Eastern Freightliner tractor to haul trailers in 1950 and in 1953, followed with its first overhead sleeper tractor, which could run on multiple varieties of fuel. Both Freightliner and Peterbilt lay claim to the first 90-degree tilt cab that eased maintenance in 1958. Along with Cummins engines, it developed a power-assist for tractors hauling double or triple trailers over high mountain passes.

Daimler-Benz AG bought Freightliner from its parent company, Consolidated Freightways, in 1981, followed by firetruck maker American LaFrance (1995); Louisville Lines; Ford's heavy-truck division Louisville Lines (1997); and Detroit Diesel (2000). Daimler/Freightliner continued to innovate, building larger sleepers and electronically assisted articulating steps. The company created a full-scale wind tunnel in 2004 to test aerodynamics and adopted its "Run Smart Predictive Cruise" in 2009 to allow modeling the road ahead to achieve fuel savings. Freightliner was chosen by the federal government in 2013 to find ways to improve Class 8 trucks' fuel economy and greenhouse gas emissions. Freightliners' Detroit engines met federal greenhouse gas emission standards in 2016—a year ahead of schedule. It was the first manufacturer to offer a suite of safety systems using radar to prevent collisions and cameras for lane-departure warnings.

As for Ford selling Louisville Lines to Daimler/Freightliner, I have a personal connection to that evolution. My cousin, John Merrifield, who grew up near Richmond, Missouri, and spent an entire career with Ford Motor Company, was head of the Ford's Louisville Lines division when Ford made the sale. The Louisville Lines division became Sterling Trucks, under Daimler/Freightliner and John Merrifield was the CEO of Sterling during its entire run as a division of Freightliner. After a few years as a standalone brand, Daimler/Freightliner rolled the Sterling division into Freightiner. Freightliner turned 80 in 2021.

Japan shows muscle in medium-duty

While heavy-duty trucks haul the goods, medium-duty trucks are the workhorses of society; they are contractor's vehicles, delivery trucks, bucket trucks and cherry pickers and farm trucks.

Isuzu has been a leading supplier of middle- and some heavy-duty trucks since it began importing into the U.S. in 1984; it's KS22 truck had an 87-hp diesel engine. Isuzu quickly became the best-selling low cab forward brand in America. Low cab forward (LCF) trucks' engines are located under the cab, rather than in front and it's easy for a user to choose the body that will work best for their application.

Isuzu has delivered more than 500,000 trucks in North America. It began assembling trucks here in the mid-'90s. Isuzu's commitment to become cleaner and more fuel-efficient has been a theme in the company's development: LCF medium-duty trucks were the first to achieve compliance with 2010 EPA regulations in 2010; 2015-16 diesel models met greenhouse gas emissions standards in 2015—a year early; and an all-electric Isuzu truck was shown at the Work Truck Show in 2018.

Japanese powerhouse Hino has been Japan's top-selling truck brand for almost 50 years, and boasts it is America's fastest-growing medium-duty truck brand but also manufactures heavy-duty trucks. Hino was established as an independent company and began selling in the U.S. in 1995; it became a subsidiary of Toyota in 2001. It has plants in West Virginia and also manufactures parts in California and Arkansas for Toyota.

The Americans: tough, rockin' and Ram-in'

Ford, Chevrolet/GMC and Ram are best known for pickup trucks, but also manufacture bigger trucks, including farm and construction workhorses such as Ford F-350 through 650, Chevy/GMC Silverado 4500 – 6500 and Ram 4500 and 5500 trucks. Ford continues to build medium-duty F-650 and 750, as well as some Class 8 F-750s.

Chevrolet began building versions of its iconic pickup in 1918 to compete with Ford's Model TT. It continues to specialize in pickups. Chevrolet and GMC trucks share architecture but have individual badges. The C/K line debuted in 1960, with pickups but also incorporating some heavier models. From 1966-2003 these General Motors twins produced the B series medium-duty "incomplete" trucks, based on the C/K chassis, that could be built out for commercial use. This was popular, adapted for use as a school bus.

The Chevy/GMC C/K series underwent several iterations until 2000. Now, its Silverado "Heavy Duty" versions, the 1500HD, 2500HD and 3500HD are capable medium-duty trucks; GMC uses

Sierra HD branding. In 2001, Chevrolet introduced the Duramax 6.6L turbodiesel, which morphed into the second-generation Duramax—producing an impressive 910 lb-ft of torque.

Stellantis (formerly Chrysler Corporation, Daimler-Chrysler, Fiat Chrysler Automobiles) Ram Trucks grew out of Dodge Trucks and became their own line in 2010, led by the flagship Ram 1500. The 2500 and 3500 were manufactured beginning in 2010, with both Hemi gasoline and Cummins diesel engines and availability as either automatic or manual transmissions. Ram also built chassis cab versions, which could be adapted for several uses beginning in 2010. Currently Ram boasts "Best-in-Class Towing Capacity," of up to 35,220 lbs., which it showed off at the 2019 Denver Auto Show, and elsewhere among other major shows and events.

The future direction of heavy-truck industry: power source technology

Along with the push for zero-emission cars by governments around the world, there have been growing demands for zero-emission trucks. That is a very tall order. Diesel engines are very efficient and pack a great deal of energy. These engines also tend to be very reliable and durable. Many diesel trucks on the road today are 20 or more years of age with rebuilt engines. Diesel fuel has powered the trucking industry for the past 90 years. Trucks have become more fuel-efficient over the years and have, for the most part, eliminated brown cloud particle emissions from their tailpipes. The other kind of emissions from motor vehicles is the clear, climate-changing CO_2.

While the industry has greatly improved efficiency and reduced CO_2 emissions, we are a long way from being emission-free. There seem to be three avenues, at this time, to reach zero-emission trucks. None will be easy nor will any be inexpensive.

Battery-electric Trucks (BETs): Just as batteries are fast becoming the preferred power source for emission-free cars, there are those who believe battery-powered semis can eventually replace the fleet of over-the-road trucks. The challenges of powering cars with batteries are magnified greatly when applied to heavy freight hauling trucks. These issues include 50 to 100 percent added cost, as well as battery size and thus weight (estimated to be 8,000 to 10,000 pounds greater than a conventional diesel at this time).

There is also the question of range, as in how far a truck can go while pulling its capacity load. Will there be charging systems en route and spaced as needed? Charging time and the amount of power required to fully recharge the batteries (current estimates are that a full BET may require 1 megawatt of power to charge it fully). Offsetting some of these fierce negatives are: anticipated lower operating costs and reduced maintenance costs over time. Freightliner, Peterbilt, Kenworth, International, Volvo, White, and Tesla are all testing ranges on early editions of BETs. Volvo has gone one step further and committed by 2040 to have a complete fossil-free product range using battery or hydrogen technologies.

Elon Musk first unveiled the Tesla Semi in 2018 and announced it would be ready for fleet use the next year. Countless trucking companies put down sizable deposits with expectations of deliveries in 2019. As of late 2023, there were still only a handful of Tesla Semis in production. As Kermit the Frog would say, "It's not easy being green!" As any car or truck manufacturer would say, it's not easy to mass produce quality products.

Hydrogen fuel cell technology (hydrogen, for short): Hydrogen-powered semis are expected to address some of the concerns of BETs. Hydrogen would take up less space and weight being carried in the payload of the truck, compared to BETs. Refueling times would be significantly less than that of BETs. A startup truck company called Nikola hailed its over-the-road hydrogen-powered semi technology. General Motors and others placed big bets on Nikola's investment in hydrogen-powered over-the-road semis. Then news broke that was no "there" there in Nikola's hype.

In an early promotion of Nikola hydrogen-powered semis, the company rolled a Nikola truck down a hill to make it look like it was powering itself. It wasn't. When word got out about the 'roll the truck down the hill lie' it led to a huge rush away from investment in Nikola and later led to the downfall of the founding CEO.

Just recently, both Toyota and Hyundai have announced major investments in hydrogen-powered technology for trucks. Both have the financial bandwidth to make their investments play out and come to fruition.

Personally, I think hydrogen makes more sense for over-the-road trucks than the gigantic batteries required for long-haul service in BETs. Building a network of either battery re-charging stations or hydrogen powered re-fueling stations for BEV powered semis will be a mammoth undertaking and cost billions of dollars. The advantage batteries have is the technology is already fast advancing for fleet of passenger cars and trucks and it's likely that will continue into the foreseeable future.

Renewable natural gas and renewable diesel-powered trucks

Compressed natural gas (CNG) makes great sense as a long-term between diesel- or gasoline-powered vehicles and the ultimate, yet extremely difficult, goal of zero-emission trucks. Various studies indicate that carbon emissions can be reduced by something between 20 and 40 percent versus traditional diesel or gasoline-fueled trucks. Additionally, compressed natural gas is home-owned and home grown. In essence, the United States has an abundance of natural gas and is actually producing more than we use, so we have become a net-exporting state when it comes to CNG. And, another bonus, natural gas is less expensive than diesel or gasoline so it buys more miles per dollar than other fossil-based fuels. Ultimately, CNG is more accessible, more affordable and easier on the environment through lower carbon emissions. CNG makes sense for truck manufacturers, for trucking companies and for the public.

Autonomous trucks

The other major, game-changing advancement in over-the-road trucks is that of self-driving technology, what some are now calling, Robot trucks. Just as passenger cars and light duty trucks get closer to driving themselves each and every year, so too do over-the-road, freight-hauling semis.

This technology, when perfected, could be game-changing. No longer will freight routes be dependent on relay system routes or sleep time for overworked drivers. If it lives up to its promise, the technology will take the pain out of engaging personnel to get the payload down the route toward the destination.

You have heard the phrase, "Don't let the perfect be the enemy of the really, really good." In this particular instance, we all must demand the perfect. The "really, really good" just isn't good enough to cut it. This technology must excel to the magnificent magnitude of perfect. When it does, the industry can literally just keep on trucking. And do so without having to rely on every mile with drivers. Organized labor, namely the Teamsters Union, has spoken out vociferously against the common use of autonomous trucks.

48

Diesel

Diesel engines remade trucking

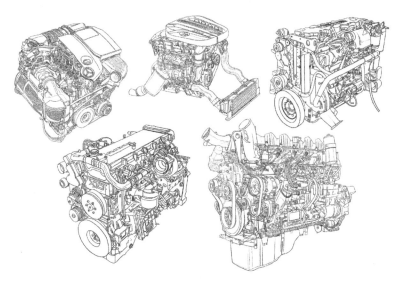

*"The diesel engine has always been my first love and I knew
it would revolutionize the trucking industry."*
—*Clessie Cummins*

Heavy-duty trucks run on heavy-duty fuel: diesel. These days, more than 95 percent of the big trucks on the road run on diesel fuel. But it took a while for diesel engines to take hold. Rudolf Diesel went from being a student of thermodynamics to inventing a new kind of engine based on the premise that higher compression created more power and efficiency. His first prototype was built in 1893 and the first successful model in 1897. It was 16.2 percent more efficient than the steam engines that were available at the time.

Other inventors' improvements only increased the diesel engine's benefits and utilities. Swiss inventor lfre Büchi added a turbocharger to the diesel engine in 1925, boosting efficiency more than 40 percent. In 1927, Robert Bosch made diesel engines even more economical and efficient with a fuel-injection pump.

Cummins

Clessie Cummins, a mechanic in Columbus, Indiana, formed a partnership in 1919 with industrialist and banker William Irwin, for whom he was a part-time chauffeur. Their first endeavor—a three-horsepower farm diesel built for Sears, Roebuck—was a bust, but with Irwin's backing, Cummins kept working to build better, more powerful diesel engines. He hit on a successful solution in 1930 that worked well in automobiles.

Truck manufacturers remained unconvinced, continuing to rely on gasoline engines. Irwin came to the rescue again, placing Cummins diesel engines in the delivery trucks in use by his California-based grocery chain of Purity Stores. The drivers liked them and the word spread. The company prospered as America's roads and highways led to more truck traffic. Cummins continued to focus on engines it sold to truck manufacturers. Cummins developed a turbodiesel in the early '50s, which increased engine horsepower by half without increasing fuel consumption. Cummins dominated the heavy-duty U.S. diesel market throughout the '50s, but the '60s brought increased competition. Cummins kept to diesels, but its attempts to scale down to the light-duty market were not particularly successful, nor were the company's efforts to diversify.

On a roller coaster driven by a fluctuating economy, Cummins increasingly turned toward non-highway engines and a growing European market. In the '80s, it faced more competition from Japan, and met it with better manufacturing procedures and cost-cutting. In the late '80s the pressure was on to meet government emission control regulations. Its offering had problems and business migrated to its more successful competitors, causing Cummins' market share to fall dramatically.

Cummins bounced back under a new partnership with Ford, Tenneco and Kubota, and its 75th year, 1993, was its most profitable ever. It went on to form several joint ventures and invested heavily in research and development. Nevertheless, Cummins and other heavy diesel engine manufacturers paid a large fine to the EPA to settle a suit that claimed computerized timing devices were used to evade emissions tests. Cummins' financial fortunes dropped with its heavy-duty truck engine sales declining precipitously. The slide continued. However, in 2001, Cummins agreed to build engines for PACCAR, the parent company for Kenworth and Peterbilt. Cummins' diversification into other sectors beyond heavy-duty diesel engines ultimately saved it. One notable development was its alliance with Chrysler in 1989, and the introduction of the first Cummins-powered Ram truck. It is a partnership that has thrived.

The Cat: Caterpillar

Caterpillar was founded by competitors Benjamin Holt and C.L. Best. Holt invented the steam tractor in the 1890s, while Best worked to perfect gasoline technology. The two tractor companies combined after World War I, to form Caterpillar Tractor. They introduced the Caterpillar Diesel Sixty Tractor in 1931 and quickly became the world's-largest producer of diesel engines. They widely powered construction, mining and agricultural vehicles. Caterpillar introduced its first truck-specific engine in 1939, but stopped building truck engines during World War II and didn't start again until the 1960s. Caterpillar introduced its first off-highway truck in 1962. By the mid-2000s, Caterpillar led the market in Class 8 engines, with almost 30 percent of the market; Cummins was nipping at its wheels. The market lead flipped in 2007, with Cummins selling more than Caterpillar.

Caterpillar ran afoul of the EPA's emission standards with its 2002-03 engines and was forced to pay a hefty fine. By 2008, Caterpillar decided that the market for truck engines was plateauing and the costs of keeping up with ever-increasing emissions standards were just too great. It made the decision to bow out as an independent engine manufacturer in North America by 2010. It formed an alliance with Navistar to continue selling trucks and engines in Asia and Europe, focusing on severe-duty trucks for construction, petroleum production and logging. There is still a robust market for "remanufactured" Caterpillar engines.

Made in Detroit

The history of Detroit Diesel goes back to 1938 and General Motors, which founded the GM Diesel Division. Heavy-duty war equipment needed lighter-weight, compact engines and GM delivered. The Series 71 performed well and gained a reputation. By the '50s GM was developing new designs and shifted their focus to over-the-road trucks. In 1957, it was the Series 53 engine, including two-, three- and four-cylinder models. They were versatile enough to be used for many industries.

GM rebranded its diesel division to Detroit Engine Division in 1965. It combined efforts with the Allison Division and officially merged the two in 1970 to become the Detroit Allison Division, which competed successfully with Cummins throughout the '70s and '80s.

The Series 60 line, introduced in 1987, included standard integrated electronics and boosted fuel efficiency. It was a huge hit and quickly became the best-selling Class 8 truck diesel engine. It attracted the attention of Roger Penske, and his Penske Corp. took majority ownership in a new endeavor, Detroit Diesel. The new company flourished and by 1993, when it was listed on the New York Stock Exchange, controlled about a third of the on-highway engine market. Its success attracted a new buyer, DaimlerChrysler. As a division of Daimler Trucks of North America, Detroit Diesel retooled and upgraded and released a new engine line, the "DD," while continuing to sell its Series 60. It sold its one-millionth Series 60 engine in 2009.

Detroit Diesel—now known simply as "Detroit"—continues to stay on the leading edge of diesel engine development with concepts such as the BlueTEC® emissions technology and Remote Diagnostics System. It also is a leading supplier of products to other industries.

The future of diesel engines

Considering the durability of diesel engines and their ability to haul huge loads, it's hard to envision that they will be replaced soon. Many diesel truck engines can rack up more than a million miles and still run. Advanced technology has boosted fuel efficiency and reduced emissions without sacrificing power.

Battery electric, as well as hydrogen fuel cell-powered, heavy trucks are beginning to surface that future watchers say will make diesel engines obsolete, but it's hard to tell if they are a real threat or just the newest shiny thing. In any event, with diesel engines powering truck fleets worldwide, not to mention all other heavy machinery across multiple industries, it will take a long time before they go the way of the dodo, even though government regulators around the globe are working hard to make diesel, as well as gasoline engines, obsolete.

PART TWO
THE PRESENT

49

Street Fight

The new battleground between car commuters and the anti-car crowd

"To build a road is so much more than just laying tarmac.
It is about providing the means for people to move freely and
safely, whether they are driving a car or riding a bicycle."
—*Unknown*

As this book documents, various means of transportation and commuting have grown exponentially over the past five decades. Vehicle miles traveled (VMTs) have grown in the United States, from 1.1 trillion in 1972, to over 3.3 trillion in 2022. And some estimates call for VMTs to, again, more than double in the next 20 years. Air travel miles are also expected to more than double over the next 30 years. Whether capacity for each of these may be hard to make grow, at a minimum, we shouldn't be reducing capacity for mobility for any common form of transportation or commuting.

Full disclosure: I'm very pro-car. I love to drive them and I've worked in and around the auto industry for most of the past 20 years and for over 30 years of my life. I drive between 20,000 and 25,000 miles each year. Also full disclosure: I also love bicycles and track my adventures as a cyclist on Strava, the popular fitness app. Over the past 20 or so years, I have cycled over 60,000 miles and I now average between 2,000 and 4,000 miles every year. I've also completed weeklong rides of 400 to 500 miles in 17 of the past 20 years. And, full disclosure: I'm very pro-aviation. I've been a licensed private pilot since age 19, 1976. And I'm a repeat 100,000 mile frequent flier in commercial aviation for 12 years running, with 3 million lifetime air miles.

I can honestly say that I'm obviously very pro-bicycle and at the same time very pro-car and pro-aviation, which I think makes the data and messaging presented in this chapter more credible.

As a lover of both bikes and cars, I don't think that it's productive, smart or logical to pit them against one another. It pains me when I see an advocacy group for either cars or bicycles present their arguments as an either/or proposition. This is not a zero-sum game. There is room for both and for other mobility options, as well. As Americans we need our full spectrum of choices.

Nevertheless, there is a fight going on, and the hyperbole accompanying it has become louder and more divisive, especially in the most tightly compressed, high-density metropolitan areas, where space is at a premium and contested by cars, bicycles, motorized scooters, buses, one-wheels, skateboards, motorcycles and other individualized modes of personal mobility.

This new street fight plays out with neighbors fighting neighbors, single-family residents fighting multifamily dwellers and bicycle commuters fighting car commuters. These fights go against the American "live and let live" ethos that has historically recognized individual preferences, lifestyles and habits. Instead, we've retreated to our corners, ready for the bell to come out fighting over the next slice of pavement.

A lot of this is based on perceptions of special treatment: If I want to bike to work, why should the city let cars drive on the same street? Flip that argument: If I want to drive to work, why should the city let my neighbor take up space on the same road for a bicycle? Why should one group be catered to over the other?

In our political universe, transportation advocacy groups get together in coalitions to lobby government—both lawmakers and influential bureaucrats—who make decisions about how to divvy up road space and how various users can access it.

Parking is an issue that's attracted a huge amount of attention, especially in the densest areas of cities. It's a multipronged issue. One controversial question is how many, or even whether, parking spaces should be required to be included in the construction of either commercial or residential buildings.

The recent trend has been for urban planners to eliminate parking minimums for residential construction in zoning codes, and in some cases even institute parking maximums. The anti-car forces—transportation consultants, bicycle advocates, and more—argue that making it harder for residents to find parking will force them to shift their mobility choices away from cars toward public transit, bicycles and walking.

But as with any choice, these decisions have consequences. Sometimes negative.

One recent example was in Fort Collins, Colorado, where the Colorado State University was building new student dormitory housing. The university reasoned that if it didn't provide parking for the students who lived there those students would leave their cars at home. It was faulty reasoning. Many of the dorm residents still brought their cars and having no place to park them close to the dorms, they filled up nearby neighborhood streets. Local residents were infuriated because it meant there was no place for them to park. The university caved to the pressure and added new parking spots on campus.

The street fight an hour south in Denver manifested in a squabble over a longstanding requirement to provide 1.7 parking spots per unit in new multifamily residential developments. The anti-car faction triumphed by turning the 1.7 spots from a minimum to a maximum, with zero parking spots possible for some specific construction projects.

The upshot was a developer pushing to the limit—and lowering construction costs—by building a 240-unit micro-apartment complex in Denver's Capitol Hill neighborhood with no parking at all! In a neighborhood notoriously short of on-street parking, the developer got huge blowback and alleviated the uproar by leasing 30 year-round spots in a nearby vacant lot.

Also in Denver, less than a decade ago, a new hotel was allowed to be built near the convention center that sported 500 new guest rooms. The hotel carried two distinct hotel brands, each with a different street accessing its front entrance and valet parking stands for those driving in. Yet no actual onsite parking sites were required nor built. Anyone driving in to check in would pass their car and keys to a valet who would literally drive the car off to a different, older hotel that had been built with excess parking. The operative term for the older hotel is "built, over-parked." The operative term for the newer hotel without parking was "built, under-parked."

Still, by filling up the neighboring hotel's excess parking, the new hotel was at the same time, compressing parking downtown, thereby making it less available and more expensive. This type of process is compounding the downtown parking dilemma and pricing many downtown cities out of attracting suburbanites and even other neighborhood city dwellers downtown. It adds to the recent trend of downtown blight, empty storefronts and offices and manifests itself in urban decay, ruining once beautiful metropolitan downtown neighborhoods.

Those are but three episodes in the ongoing battles over parking. There are others going on in the downtown areas of many cities. City and transportation planners' basic calculus is similar to what the planners at Colorado State University thought: When parking isn't readily available, people will leave their cars at home when they go downtown, local businesses will attract those same people to shop and everyone will be happy. That isn't what seems to be happening.

The unintended consequence of radically limiting parking is to discourage people from going downtown at all. Instead, they stay in their own neighborhoods or often stay in the suburbs and patronize businesses that welcome them with or without their cars. Without customers, the downtown businesses sputter and eventually close while suburban businesses thrive. It's pretty basic economics that is often missed by urban planners. Or maybe these economics are overlooked intentionally.

What's the logical driver response?

Drivers do rightfully get upset over clogged roads and lanes restricted or taken over for bicycles, scooters and buses. Nobody likes to be inconvenienced, but there are some valid reasons. When cars aren't handled alertly and safely, they can become killing machines. The data is pretty clear. Traffic fatalities in the U.S. reached a high of more than 52,000 in 1972. For the next 40 years, there were annual declines, so much so that by 2014 the toll had declined to about 32,000. It should also be noted that while traffic fatalities dropped by more than a third over those years, vehicle miles traveled (VMTs) increased by more than 300 percent, making the decline in fatalities even more significant.

Specifically, in 1972 (the year seatbelts were required by the federal government to be installed by manufacturers of new motor vehicles) there were 1.1 trillion VMTs in the United States. By the year 2019 (most recent data found for this publication), the number of VMTs in the US had increased by 300 percent to 3.3 trillion. Now in 2024, its safe to say we have more than tripled VMTs since 1972, the year seatbelts were required in new cars.

Since the modern-day low point in 2014, traffic fatalities have surged recently to alarmingly near-record levels. That should be a warning to all of us whether we drive cars or ride bikes that we have a responsibility to pay attention to the road, avoiding distractions and anything else that could remove our eyes from navigating the road ahead safely. As drivers, at some point—me included—we find ourselves driving in a way that can be viewed as driving distracted. It's very concerning that many fatalities involve cars and bicycles or cars and pedestrians.

Ironically, while fatalities have risen, vehicles have become much safer, providing much better protection for the occupants. Seat belts, airbags and better interior and exterior design make it much less likely, except in extreme head-on and rollover accidents, that vehicle occupants will be seriously injured or die.

Safety equipment is being designed and becoming more prevalent to prevent cars from injuring bicyclists and pedestrians, but it will take several years before the entire fleet of cars and trucks is equipped with these extremely effective safety systems. In the meantime, most fatal or injurious bicycle and pedestrian encounters with cars are happening in major cities that have been most aggressively removing traffic lanes and converting them to other uses.

Obviously, drivers need to be much more mindful and vigilant, and avoid erratic or aggressive driving that increases the risk of traffic injuries or fatalities. Law enforcement needs to do more to crack down on aggressive and dangerous drivers. Car owners need to keep vehicles and their safety equipment in working order, and wear seat belts. Children should be buckled into age-appropriate and properly installed car seats.

What many cycling and pedestrian advocates want

There is a growing subset of city dwellers—especially those in the densest urban centers—who enjoy their own mobility freedom by using bicycles, motorcycles, scooters, one-wheels, mass transit or walking. More power to them!

However, some of these folks have become radically reactionary to cars. For them, it's not just about making it harder for people to buy, own, park and drive cars in urban areas. Some urbanites want to ban cars outright—at least in city centers. For those, less power to them!!

Who comprise the advocacy groups? They are self-avowed "urbanists." I believe they are essentially hyperlocal, for whom a long trip may be 15 blocks to the grocery store or 30 blocks to a doctor's appointment. They seldom cross from one side of their city to the other, let alone travel across their state. Many advocate what they call "15-minute cities" and the benefits of being able to reach all essential services within 15 minutes or less. They rarely seem to consider the problems faced by people in our society who by necessity (expensive housing, schools, employment) have to commute 60 minutes or more to jobs or whose jobs demand they drive thousands of miles each year.

These anti-car urbanists often identify as pedestrian advocates, bicycle advocates, transit advocates, safe street advocates, urban transportation planners and climate activists. But they all seem to agree that cars are an existential threat to a happy existence to them within their city.

Take a listen to a podcast originating in New York City titled "The War on Cars." Stickers proclaiming the "War on Cars" and "Cars Ruin Cities" are cropping up at inner-city cycling events. The urbanist agenda for making streets over for bikes and pedestrians includes:

- Protected bike lanes from cars and other traffic with raised dividers, posts, bollards or other permanent barriers.
- Diverting traffic by putting up barriers (appropriately called diverters) that restrict local street access only to bikes, pedestrians and other non-vehicular users.
- Reducing speed limits, especially on neighborhood streets. Streets with 35 mph speed limits are now being slowed to 30 mph; 25 mph speed limits are being slowed to 20 mph. This has happened recently in Denver and other cities across the country.
- Eliminating right turn on red after stop at traffic intersections. Cyclists call this "the right hook," created when drivers fail to look right after stopping and then enter an intersection to make a right turn. Some large cities have already acted on this change and others are being lobbied to do so.
- Adopting rotary circles or roundabouts, already common in Europe as a traffic management system. Traffic coming from any direction is required to slow down in order to enter, thereby "calming" traffic. Advocates are pressing more US cities to adopt them. Interestingly, some cities are installing traffic circles on neighborhood streets where some residents say they really aren't needed or wanted.
- Adding speed humps/bumps to force cars to slow down to a crawl or scrape their undersides. These are showing up in countless neighborhoods across the US.
- Eliminating laws against jaywalking, the process of crossing streets in places without the use of a crosswalk. These laws date to the early 1900s when policymakers saw the need for clarity of right of way when cars were first gaining prominence and pedestrians would otherwise cross a street at any point available.

- Closing streets to cars entirely, especially in downtown areas. It's being advocated more and more often.
- Removing car lanes from primary commuter corridors, thereby reducing road capacity, called a "road diet," increasing traffic congestion and creating longer drive times and greater frustration for commuters.
- Advocating against expansion of lanes due to growth in population or alleged overuse of existing roadways.
- Eliminating or drastically reducing parking minimums, which will save developers money. Advocates claim it will make new housing more affordable.
- Reducing public parking spaces, which will drive up costs in nearby parking lots and discourage people from driving and perhaps (they hope) even persuade them to use public transit.

Where the Public stands

Each of these anti-car restrictions seems to spark its own hyperlocal fight. But among neighborhoods that are underserved by transit and/or are inconvenient to access by bicycle, sentiment often swings toward keeping easier access to cars.

In the run-up to the Covid-19 pandemic shutdowns, a public opinion poll of voting-age residents was released and was made public at the time. It included both statewide Colorado and Denver-only questions. The poll of registered voters turned up some interesting and compelling results:

- 89 percent of respondents statewide in Colorado reported they owned or leased a car.
- 88 percent of Denver residents reported owning or leasing a car—statistically the same as respondents statewide.

- By a ratio of three to two, respondents opposed governments intentionally closing lanes in streets to vehicles for the express purpose of making it harder to commute by car.
- By a ratio of three to two, respondents opposed governments intentionally eliminating public parking spots for the express purpose of making it harder to commute by car.
- When asked how respondents normally commute to work, a vast majority (more than four to one) drive to work vs. the next-favored choice, public transit.
- When asked if they ever chose not to drive to downtown Denver because of lack of lanes and or parking, a large majority reported that restrictive steps the city has made to get there led them to cancel plans to visit downtown Denver.

Denver's transportation planners chose an unusual lane-closure project on South Broadway, a major commuter thoroughfare near downtown. I was instrumental in organizing a rush-hour traffic count survey to assess the modes of transportation commuters used. The counts took place in October 2019 and at approximately the same month and time, subsequently, in 2020, 2021, 2022 and 2023. Prior to 2019, Denver had reallocated one full lane from cars to bus-only. About the same time, another traffic lane for cars was reallocated to bicycles. Together, these re-allocated traffic lanes reduced vehicle capacity on the busy corridor by 40 percent, from five lanes to three.

A small group of volunteers counted commuters—not vehicles —and what we discovered was jaw-dropping. In 2019 between 4-6 p.m., with a temperature in the low 30s:

- A whopping 89 percent of over 6,000 commuters commuted in cars and were compressed into three traffic lanes.
- Only 11 percent of commuters traveled by bus.

- Fewer than 1 percent (0.2%) traveled by bicycle and of those, only five of 11 cyclists actually used the protected bike lane. Another five cyclists, instead, used the Bus Only lane and one rode in the middle of the three traffic lanes (wearing no helmet and without any bicycle lights).

In 2020, also at rush hour, with the temperature in the low 60s:

- 97 percent of a total of over 5,000 commuters travelled in cars.
- 2.7 percent commuted by bus.
- Nine (9) were on bicycles; only two actually used the protected bike lane, three rode on the sidewalk and four used the Bus Only Lane.

Needless to say, when the results were released, a lot of people who were advocating to make it even harder for cars to commute through the corridor were nonplussed or upset. These figures were not part of a scientific study, to be sure, but they are very indicative of how the city's efforts to change commuters from cars to transit or bicycles didn't seem to have much impact, even when the weather was cooperative.

Is this all to say cyclists should stay away from the area? Not at all. Neighborhood streets adjacent to this South Broadway business district are the self-selected and logical choice for those of us who want to visit South Broadway by bicycle.

I was also involved in another informal count of bicycle use in a residential area where the city had removed 300+ roadside parking spaces to install bike lanes in both directions, over the protests of the neighbors. Monitoring traffic from 6 a.m. to 6 p.m. on a Thursday in early-cool-temperature-trending October revealed that the vehicle breakdown included 2,189 cars and only 51 bicycles. In other words, on an already very safe neighborhood street, with no known or re-corded fatalities in the past 20 years, 300+ roadside parking spaces were eliminated and resulted in 51 cyclists utilizing the route, even though most or all would have already been safely using the route, even while street-side car parking was still allowed.

Granted, even though these surveys were not necessarily scientific, they do indicate that transportation planners and city residents aren't seeing eye-to-eye on transportation utilization. The battle lines have been drawn and the fights are continuing in communities around the country. Social engineering of transportation and mobility is well advancing, even when it is the wrong thing to do. Often the people whose lives will be affected most don't anticipate these battles and are ill-prepared to fight back. Their opportunity will come. Indeed, in my opinion, the opportunity is fast approaching.

50

Vans

How "Road trip!" has evolved to "Van life"

The beauty of van life is the freedom it offers to explore new places, connect with nature, and live a simpler, more fulfilling life on the road and beyond. We can be reminded that the best things in life aren't things at all, but experiences and memories made along the way."
—Phil Ingrassia, Recreation Vehicle Dealers Association (RVDA)

Whether by choice or by necessity, traveling, sleeping and even living fulltime in vans have become part of the fabric of American life. Social media abounds with stories and photos. The Instagram hashtag #vanlife has accumulated more than 18 million posts. A quick search on the Internet turns up websites devoted to the lifestyle: Projectvanlife.com, gnomadhome.com, Kombilife.com, faroutride.com, justvanlife.com, go-van.com—and on and on. There are many apps designed to connect Van lifers with advice, campsite locations and each other.

Van life in the United States encompasses a wide spectrum of Americans. There are impoverished retirees and young people fleeing the lack of affordable housing and who live in used vans they've converted themselves. But there's also been a boom in new and newer Mercedes-Benz Sprinters, Ford Transits and Ram ProMasters that can be seen on highways returning from weekend recreational jaunts or longer vacations.

The recreational vehicle industry has enjoyed very good years, more than partly due to the COVID-19 pandemic. The Recreational Vehicle Industry Association (RVIA) reports that in 2021 (most recent data) manufacturers shipped more than 600,000 RVs of all types, including towables—trailers, popups and fifth wheels. Motorhome sales in 2021 increased by 37.8 percent but by far the fastest-growing segment was Class B campervans, which increased 91.5 percent year over year.

Van life hotspot

Not surprisingly, Colorado's scenic wonders and outdoor activities have made it a natural hub for Van life.

A survey of 725 Van lifers by online publication *Outbound Living*, ranked Colorado #2, behind California among Van life states.

A more recent survey by the RVIA looked at Class B campervan demographics, which, much like Colorado's, trend younger. Young families accounted for 42 percent, millennials and Gen-Z accounted for 45 percent. Two-thirds are male and more than half have no children at home. They like outdoor sports (44 percent), fishing (32 percent) and water recreation (32 percent). Their motivation for owning a campervan, according to the RVIA survey, include "maintaining control over one's own itinerary, spending time outdoors, and visiting locations with natural beauty."

The younger cohort of Van lifers may think they've stumbled onto something new and different and romantic—whether they're in a do-it-yourself (DIY) rig or a tricked-out campervan. Some of the photos on van life Instagram and Facebook pages demonstrate some of the romance and adventure they feel as they are able to pursue remote employment while shifting their location from ocean to mountains to desert and back. But Van life can actually be traced back years—even centuries.

A short history of vans

Arguably, van life began with the Romani ("gypsies"). There are accounts beginning in Medieval times that these nomads traveled throughout Europe in horse-drawn wagons. A 14th century monk wrote that they "rarely or never stop in one place for more than 30 days." They were persecuted wherever they went.

Motorized campervans made their American debut early in the 20th century. The 1910 Pierce-Arrow Touring Landau with a fold-down bed and sink was displayed at the Madison Square Garden auto show. And "The Vagabonds"—Thomas Edison, Henry Ford, John Burroughs and Harvey Firestone—outfitted a Lincoln truck for annual camping trips between 1913-1924.

Fast forward to 1950. Volkswagen began producing a box on the chassis of its Beetle, the Type 2. The VW Microbus was useful as a commercial vehicle, but people almost immediately saw that the interior could be adapted for other uses, including camping. Volkswagen partnered with Westfalia to produce conversion kits, and by 1956 the VW Westfalia camper had come to America. Though low on horsepower, it was high on appeal, and it became a symbol of a culture outside the mainstream of American life, earning the nickname "hippie bus," although plenty of non-hippies saw its attraction, as well. Many are still in use today.

Volkswagen discontinued the Microbus in 2014, but has announced an all-electric version, the ID. Buzz, that it plans to introduce for the 2024 model year.

American-made commercial vans had more power and room than the VW, and adapted well to camping. Tricked out with beds, carpet and often stoves and refrigerators, they continued to acquire fancier accessories like mood lighting and sound systems, more luxurious seating and sleeping arrangements and distinctive paint jobs. Many of these are still on the road, as well.

A happy medium: the class B campervan

Over the years, the term "recreational vehicle" (RV) has embraced everything from buses to semi-trailer trucks that have been converted into homes on wheels (Class A) to motorhomes with a bed over the cab and conveniences like built in showers and toilets (Class C). In between is the Class B van, which has gained huge popularity in the last few years because it is large enough to accommodate multiple beds and other amenities, but smaller, more efficient and easily maneuverable.

The Class B campervan was developed beginning in the mid '70s in Canada, by Roadtrek and Pleasure-Way. According to Phil Ingrassia, president of the Recreational Vehicles Dealers Association (RVDA). "They kept the flame alive by outfitting vans into real RVs, with kitchens, bathrooms and sleeping that were above and beyond what the early van campers could've imagined." But it was Mercedes-Benz and its Sprinter that made it possible for an adult to stand up inside.

Sprinters brought more height and a narrower wheelbase to the industry and gradually, the Mercedes-Benz "Eurovan" silhouette has become the U.S. standard. Ford brought its Transit stateside, and RAM (part of Stellantis, formerly Fiat-Chrysler) modified its existing vans to European-style standards, dubbed the ProMaster. These three brands dominate the Class B market.

Familiar names in the RV industry including giants like Winnebago and Airstream build far-from-basic campervans based on Sprinters and ProMasters. There are also scores of smaller van conversion companies. Many of them will advise and sell van parts and accessories to DIYers, like solar panels and additional water storage, but their business primarily is building and selling completed units or converting van shells that clients have acquired and brought to them for conversion.

Campervan entrepreneurs

Matt Felser, co-owner of Dave & Matt Campervans, and Eric Miller, co-owner of Tourig, could be considered poster boys for van life. They are van enthusiasts, van converters who have lived full or part-time in their Class B campervans.

Matt and his business partner, Dave Ramsay are living the van life dream. They own a successful van conversion business based in Gypsum, Colorado. The partners graduated from Williams College and went their separate ways – Dave to a hedge fund in New York and Matt to being a ski bum. Eventually, Dave quit finance, converted his first van, and then started a small van rental company.

Matt lived in Tahoe City, California before migrating to Vail, where he taught Spanish. "I was exploring my next vehicle to bike and ski outside of school," he said. In 2016-17, "The only way to get a campervan was to get a Winnebago or Roadtrek or get a custom van for nearly $100,000. Not doable! I had people who talked me into I could do this myself." He found a used ProMaster in Texas, "watched about a thousand YouTube videos, everything from flooring to electrical to even how to use power tools." Dave arrived and about four months later, the "labor of love" was finished.

They took it throughout the U.S. in 2018. "The turning moment was the amount of people on the trip who wanted to check it out," Matt said. They reasoned that if Matt could do it on a teacher's budget, it was a viable business concept purchasing used vans and converting them. "We sold the one, then two, then four. We brought on some friends and now have built 350-plus in the last three years," Matt says.

Dave and Matt Campervans work mostly on Ram ProMasters; they are licensed Ram dealers, which allows them better access to inventory, although during the pandemic of the 190 orders they placed they only received 19. Matt says the ProMaster's front-wheel drive was important for better handling in adverse conditions. It's reliable. It also has the most interior space. They are exploring diversifying into Ford Transits.

From the beginning, their model has been to "provide everything you need and nothing you don't." It makes it possible for them to keep their prices lower while offering maximum design flexibility to their clients.

The company has just bought a new 39,000-square-foot facility in Rifle and, once the vehicles start flowing again, their goal is to build 1,600 vans by the end of 2024. "We want to be the largest private manufacturer of RVs in the country and we're proud to do it in the state of Colorado," Matt says.

Dave and Matt offers 10 customization options in three lengths. "Efficiency keeps the price point down and keeps more people on the road. "Everything we do removes the bells and whistles and replaces them with a lot of flexibility and expansion," Matt says. Prices range between $65,000 and $76,000.

"We engineer vehicles that are open enough to take in all kinds," Matt says. "Most sales are to Millennials, followed by Boomers. The most interest across social media is Gen Z. The most prominent in the industry as a whole are Gen Xers. It's really cool to see if it's a Millennial entrepreneurial-minded who lives full- or part-time in the vehicle or a Boomer who wants to see the country or a Gen Z—a lot of grad students who want to reduce their overhead and don't want to live in a dorm anymore."

As for Dave and Matt, they've both lived full-time in vans. Dave still does when he's in Gypsum, although he keeps an apartment in Denver. Matt Felser's van has been his only vehicle for the last four years. He lived in it full-time for two summers and "still uses it as much as I can."

The high end of Van life

Tourig, based in Golden, Colorado, is at the other end of the price—and luxury—scale. Co-founder and CEO Eric Miller, says the idea for Tourig grew out of his job as a traveling sales rep in the outdoor industry. "I was spending a lot of nights in hotels and in tents and thought there's got to be a more efficient way to do this." For eight years before his daughter was born in 2009, Eric spent 150 nights a year in one of two vans he and a partner converted. "It was fun to watch people's eyes light up when I would pull in and get out with my dog and nobody really cared about why I was there … they cared about the van and the dog," Miller says.

The idea kept percolating, but it wasn't until 2014 that Eric thought it was time to start a business. He called his partner, Paul Bulger, an experienced marine outfitter and skillful carpenter. "Coming from the sailing background he understood what it was like to travel in a confined space … it's the best part of what made us what we are today because of his quality and attention to detail," Eric says.

In 2015, they built their first van in a 1,000-square-foot building in Nederland, Colorado. The business took off. According to Eric, "All of a sudden, the phone rang and somebody said they wanted one too. Then it rang again and before you knew it, we had people lined up and needed to hire some staff and off we went." The first year it was just two men converting two vans. The second year it was 10-12 vans. Tourig's business model now calls for about 50 per year. "It allows us to really manage our supply chain, keep quality consistent and always elevated, and it makes it exclusive," Eric says. Exclusive means prices range between $225,000 and $300,000 for a finished new Mercedes-Benz Sprinter.

Tourig works exclusively on Sprinters, and the company has a dealership license that allows them to source new Sprinters directly from the manufacturer and sell used. Tourig branched into Ford Transits in 2021 because Ford can produce and deliver vehicles that can be serviced at thousands of dealers across the country. There are far fewer dealers who can work on a Sprinter. That is partly why Tourig has doubled its space to include a service facility where they can work on vans, not necessarily Tourig-built vans. It also provides more production capacity.

A full van conversion takes Tourig about six weeks to complete. That includes about four weeks of production time plus two weeks of quality control and final detail. "We're different from an RV company. They don't produce at our level. A Class B RV retails for $180,000 and they might produce 30 a week. Our guys are artisans. They're craftsmen and it's never enough, sometimes to our detriment," Eric says.

While Tourig produces on the high end, Eric has respect for what more mass-market manufacturers offer. "I think the RV companies do an amazing job of giving people a lot of stuff for a compelling price. What we provide is an experience." The Tourig experience clearly must be worth waiting for, since they have clients waiting and delivery is 12-14 months in the future.

As for as the Tourig client? "We cater to 40-45 at the low end—family, established, job, means to afford. We're seeing a lot of people in their 60s and 70s who a few years ago would have gone to a Winnebago or Airstream because that's what you know. Now those people are saying, 'hey! I still want to be cool. I want to be relevant and out there … Why should I settle for a mass-produced product?'" Eric says.

"Hard to explain to somebody how it changes your life because, unlike a tow-behind camper or a big RV, vans have an attraction to people. Go back to the '60s, right? Vans were cool. Now they've become so sophisticated and luxurious that anybody can participate. Anybody from the people who are just getting started in van life up to vans that are doctors and lawyers who have a little more discretionary income."

Eric agrees the pandemic has contributed to the expansion of the business, but says he started to see an uptick in the industry in 2018. He couldn't afford a manufactured van when he first wanted one, so he built his own. "Now, today there are 200 companies in our little niche if you just talk about one guy in a shop to 100 guys in the back. We're in-between that with about 40." And that's just in the last three to four years.

Eric Miller has owned 10 vans over the years and his current Sprinter—his daily vehicle—is his favorite. "Had a successful career and a good life—all things essential to the American Dream. I'm happiest with my van, and whatever I can fit into it going down the road."

At the other end of the spectrum from the boutique upfitters is Summit Bodyworks, a part of Transwest Automotive Group, a dealer for new and used RVs, trucks and trailers. Summit upfits commercial vehicles for its national clientele. Need a bookmobile or bloodmobile? Summit Bodyworks is the place to go. Although Transwest already sold several mass-produced RV brands, Summit jumped into upfitting Class B campervans in 2019, even before Covid juiced the market.

"We upfit all other vehicles, so why not make that bridge? The Class B market is out of control and continues to rise," says Summit's CEO, Meredith Lyons. She believes Summit's Antero brand of Class B vans will continue to prosper, even once the pandemic is officially over (the pandemic officially ended in May of 2023). "The world has taken a different look for how to vacation," she says. "Once people see that they can sleep in their own sheets and have their own things, they see it's a nice way to travel."

Lyons says some of her customers already own large RVs and some who use their vans as a hybrid: part RV and part commuter vehicle. "People hauling kids around. Skiers who can get dressed without hauling everything into the lodge and don't have to go through the lunch line." She had one doctor who would sleep in it and could even meet patients in it.

Working out of two buildings in Fort Lupton, Lyons' team of 15 or so, including the eight-employee production crew, turns out seven units a month. That's if she can get enough Ford and Mercedes-Benz chassis. Recently, Summit has added Freightliner to its Class B chassis mix. "Last summer I still had chassis outside waiting. Now we get too close to comfort waiting—I would like to have two outside waiting [to be worked on] at all times, she says. "Supply chains are getting better but we're a ways from saying, 'Oh, they're good.'"

Lyons involves her entire team in making decisions because they represent the customers and the producers and they are constantly looking at changes in models, colors and configurations, within the reduced bounds of the supplies she can get. "You can't just sit and think it won't change," she said. "We're always looking for a way to make it better. If you stop and rest, you're going to get left behind."

Who are Van lifers?

The van life movement—people view it in those terms—actually is less about full-time living and more about part-time enjoyment. In the RVDA survey, only 1.5 percent of total RVers live in their vehicles. More than a third are over 55, they are 70 percent female, and the average annual income is under $65,000. The top full-time living RV choices are trailers, fifth wheels and larger motorhomes.

Class B campervans represent only four percent of RVs, and their demographic is quite different. The average income for 65 percent is more than $65,000. Owners are more likely to be male (66 percent). The overall age breakdown is 51 percent between 18-54 (young families, millennials and Gen-Zers) and 49 percent over 55. Almost 59 percent have no children in the home.

"While the numbers are not huge, compared to other RV types, there's a certain acceleration of the van market because it offers a lot of things for buyers: viability, flexibility and the chance to go places where you couldn't go and stay if you didn't have these types of vehicles," says RVDA's Phil Ingrassia. And with the pandemic, people were "saying they didn't even think of an RV until their [travel] options were limited. For some people the van camper was a perfect … also a prestige … way.

"These are not entry-level campers … They are a premium purchase for a lot of folks. But the economics of higher-end vacations are such that if you're going to use it for a certain number of years, a lot of people are doing the trade-off and opting for vans," Ingrassia points out.

Living the part-time Van life

Travis Berry is a 56-year-old Denver resident who bought a four-wheel-drive Sprinter four years ago. It's equipped with a bed and a refrigerator. "Pretty spartan version. For my purposes it is perfect. It can get anywhere—small enough and inconspicuous. You can park it anywhere. Has the comforts I need. Kind of a mobile biking, camping, skiing headquarters," he says. He admits he "lusted after one of those old VW campervans for years." Travis never followed through until he saw Sprinters. "It was all of that plus—new with safety stuff and four-wheel-drive and not something I need to worry about breaking down on me," he says.

Travis has taken his campervan around the West, driving it to Wyoming for the 2017 solar eclipse. The longest, including several cycle trips around Colorado roads, was the four-day Wyoming trip. What he's found is that van owners are "a total tribe. I'm floored at the explosion of them. I wanted one for a long time and they were sort of rare but now if you go to the mountains or ski areas or to a trail, they're all over the place."

He admits his wife isn't as in love with his Sprinter. "One Valentine's Day I did get a portable toilet that hooks up to the floor, but I don't use it much." And he sees that he might want to upgrade in the future, "… when I'm done working or when I start to slow down and spend more time in it—getting one with more creature comforts," including a better bathroom for his wife.

Phil Hayes went a different route. The cost to purchase a fully converted van was daunting, so he looked for a used, low-mileage Sprinter, which he found in Omaha. He worked with VanWorks in Fort Collins, Colorado after having done some prep work on the vehicle himself. "We ended up with a fully converted van for about $65,000 to $70,000, and it could have cost us twice that," he explains.

It's definitely a family affair. He and his wife have a 17-year-old son and a 14-year-old daughter. "It's a little tight. Sometimes we put the kids in a tent." His Sprinter has a bed platform and a "garage" for storing gear.

"We have really enjoyed it. We might end up selling and upgrading to something nicer or use the proceeds to buy a cabin. We don't have enough time to just drive around," Phil says. He's taken the Sprinter on camping trips including a mini tour of Colorado during the pandemic, to Kansas City for a Broncos game and to at least a dozen music festivals. In fact, he's shrink-wrapped the Sprinter "like the Grateful Dead in red and white with a lightning bolt on the front."

Beverly Razon and her husband have a 170-inch Sprinter. "We'd been wanting to buy and looking to van experiences—probably around 2019 before the whole Covid thing and didn't pull the trigger," she says. Other expenses and a desire to do some international travel for their family got in the way. A road trip to the Northwest in a regular RV persuaded Beverly and her husband that a purchase would be worthwhile. "The kids loved it, the dogs loved it," she said.

They had to go to Kansas to find their Sprinter and used an outfitter in Salt Lake City since the demand for custom vans in Colorado was so great. With two children, now eight and 10, they needed four seats, two beds, a kitchen and space for some gear and their aging dog.

Razon's Sprinter has been on some long hauls in the year they've owned it. With family on the East Coast and fears about flying during the pandemic, they traveled in their van instead. They've visited national parks, caravanning with some neighbors. "We didn't have to go in anywhere except restrooms along the way," Beverly enthuses. The ability to fix all their meals is a plus. "It gave us flexibility to adapt to any situation and do any excursion. We also enjoyed the fact that the kids have space."

Their Sprinter included an extra water tank that allows them to hook up an outside shower and clean off dirty gear before stowing it. They haven't purchased a toilet, but it's in the works. "We don't want to pamper the kids that much," she jokes.

Having the ability to work while on the road also has been a game-changer, Beverly acknowledges. "To me, it's checked so many boxes about being able to travel and have the comfort of our own personal space."

Beverly is no stranger to Van life. Growing up, her family had a conversion van and went through the Grand Tetons, southern Oregon, the California redwoods and coastline. She says, "You either love it or you hate it. Mostly it's about creating memories and experiences that we may not otherwise have."

The future is just over the next hill

When the RV industry is looking at the future, just like the rest of motorized transportation, it looks electrified. Many vans are already equipped with solar panels for heat and light. Full electrification is coming fast.

"It's almost a given that the van campers will be the first (recreational) vehicles that are run on an electric chassis," Ingrassia says. He's hearing that from the largest manufacturers like Winnebago, which has already unveiled its e-RV concept vehicle with a 125-mile range that can charge "almost anywhere" in 45 minutes. According to Jamie Sorenson, Winnebago's director of advanced technology, the e-RV is "a new solution for people to explore the outdoors with a smaller environmental impact."

Looking down the road, Ingrassia thinks the future is very bright. "There's a lot of potential for the adventure van market, especially as people take a look at the features these newer vans have. A whole new contingent of people interested in EV vans will be leading."

51

RVs

**Recreational Vehicles (RVs) –
rooted in history – more popular than ever**

*Recreation Vehicles (RVs) are a symbol of freedom and adventure,
allowing adventure seekers to explore the open road and create
lasting memories with loved ones. RVs have become an integral
part of American culture, representing the desire for
exploration and the pursuit of new experiences.*
—Jon Ferrando, Founder and CEO of Blue Compass RV

If you think about it, the first recreational vehicles (RVs)—ones
that could be "lived" in, or at least slept in—really belonged to
snails and hermit crabs. Perhaps it was from observing these
examples from nature that humans figured out they could move a
dwelling unit around fairly easily.

While the history of the modern RV in the U.S. began in 1904, these wheeled or towed vehicles actually date back much further in history. Think about the Gypsy wagons in medieval times. The people we now call Roma—then called Gypsies because they were thought to originate in Egypt—traveled throughout Europe, living in horse-drawn wagon caravans. Wherever these wagons stopped they seemed to bring thefts, pickpocketing and forbidden acts of magic. Local authorities would move in and the vagabonds would move on.

Then, there was the traveling circus. The first recorded example in the U.S. was John Bill Ricketts' show, which toured the East and Canada in the late 1700s. The traveling circus's heyday was in the mid-1800s, taking long caravans of gaily decorated wagons holding performers and animals throughout the U.S.

Rolling west in covered wagons

An American version of the Gypsy wagon was the covered wagon. The covered wagon was modeled on the Conestoga wagon used by freight haulers in the eastern part of America, but smaller. Their greased, canvas-covered tops, seen from afar, resembled sails on the horizon, thus the nickname "Prairie Schooners." They moved settlers westward, beginning with the opening of the Santa Fe Trail in the 1820s, and increasing with migration along the Oregon Trail. One major wagon train of 120 wagons and about 1,200 settlers left from Missouri on the Oregon Trail—"The Great Emigration of 1843."

It took settlers four to six months until they arrived at their destination, carrying all their supplies in the wagons, which weighed up to 2,500 pounds. Generally, the wagons weren't used for sleeping—tents were usually assembled nightly and were struck each morning, or people slept beneath their wagons. In the event of bad weather or an Indian attack, wagons offered shelter. Arriving at their destinations, settlers could live out of their wagons—the 19th century version of today's RVs.

The birth of the modern RV

A manually constructed shelter built onto a vehicle in 1904 is credited as the first RV. According to a report in *The Smithsonian Magazine* (Sept. 4, 2018), it had incandescent lighting, an icebox, a radio and slept four. But the earliest vehicle that most resembled today's RVs was unveiled at the 1910 Madison Square Garden auto show: a Pierce-Arrow Touring Landau with a fold-down backseat that converted to a bed and fold-away sink that increased space.

"Tent trailers," moderately priced and meant to be towed, carrying tents, sleeping bags and the other accoutrements needed for camping appeared in the 1910s. They developed to include a collapsible tent, cots and storage for cooking equipment.

The Conklin family headed out from New York in 1915 on a camping trip across America in its "Gypsy Van," a 25-foot vehicle built by the Gas-Electric Motor Bus Company. Conklin's house-on-wheels with its electric lights, kitchen, built-in furniture including beds, entranced the media. *The New York Times* marveled it had "all the conveniences of a country house, plus the advantages of unrestricted mobility and independence of schedule." Ransom Eli Olds introduced the REO "Speed Wagon Bungalow" in the mid-1920s, about the same time as the Hudson-Essex "Pullman Coach."

Americans had discovered the joys of the Great Outdoors and recreational camping: the "Back to Nature" movement. In fact, a group of famous men dubbed the Vagabonds—with names like Thomas Edison, Harvey Firestone and Henry Ford—took annual camping trips between 1913 and 1924 in a customized Lincoln truck. In the late '20s and early '30s, Americans again headed west, living out of their vehicles—the migration brought on by the Great Depression.

The covered wagon – motorized

Pharmaceutical executive Arthur G. Sherman had a bad camping experience while trying to erect a tent on a trailer, and his family got soaked. He hired a carpenter to build his own version of a covered camping trailer. Sherman's "Covered Wagon" was displayed at the Detroit Auto Show in early 1930. It was six-by-nine feet, had windows on the sides and two in front, and included domestic amenities like built-in furniture and storage. Expensive at $400, Sherman still sold 118 of them, grossing $3 million by 1936.

In 1929, Wally Byam repurposed a Ford Model T chassis with a teardrop-shaped structure and built the first Airstream trailer—not yet the streamlined stainless steel-clad version we know now. It sold for $500 and up and was light enough to be towed by an average car.

RVs were deployed as mobile hospitals, morgues and jails during World War II. The military bought thousands for enlisted housing. But when the war ended, returning soldiers—perhaps tired of "camping out" in foxholes—decided a better experience was desirable, and the RV industry happily obliged. Among the offerings in 1952 was a 10-wheel luxury motorhome: carpeted, with two bathrooms, television and even a swimming pool. Its $75,000 price amounts to $810,000 now—not so different from some of the best 2021 RVs.

The era of the modern motorhome/recreational vehicle had begun. And the establishment of the Interstate highway system made it even more attractive.

Many types to choose from

The RVIA estimates that 12 percent of RVs sold in 2019 in the US were motorized:

- Class A is essentially a repurposed bus with a flat front end, including the top-of-the-line Canadian Prevost costing more than $1 million, and the Georgetown and Winnebago for more than $100,000.

- Class B is the repurposed van. Mercedes-Benz's Sprinter, Ford Transit and Winnebago's Rebel are examples.
- Class C includes RVs that generally are built on a pickup or van chassis with the structure coming over the cockpit. There is a "Super C" class built on a semitrailer chassis.

Side trip to microbuses and conversions

Who can forget the first Volkswagen microbus, the Typ (Type) 2, beloved of youth during the freedom-loving days of the '60s and '70s? According to Volkswagen, the original design was sketched out on a napkin in 1947 by Dutchman Ben Pon Sr., who imported it into the U.S. beginning in 1950.

"It took you everywhere with your friends, it was a car but also a home on wheels, it was both reliable and unconventional, it was highly emotional," according to VW Chairman Herbert Diess, speaking in Pebble Beach, California, while introducing a concept all-electric VW microbus, dubbed the ID. Buzz, planned for production in 2025. It looks amazingly similar to the iconic microbus.

A generation of microbus imitators followed and eventually morphed into the conversion van, which appeared in the 1970s. The mostly standard commercial utility vans were customized with added seats, carpeting and other features to provide the creature comforts Americans wanted while they camped out in style. The vans are still popular and are part of Class B.

Lots of trailers, too

The RVIA estimates that towables comprise 88 percent of RVs on the road. They also are categorized as:

- Pop-up campers with folding sides and tops.

- Travel trailers of 13-40 feet, with many floorplans and options. Often with pop-outs on the side.
- Fifth wheels of 25-40 feet, that hang over and are attached on the bed of a pickup. Also with lots of floorplans and options.
- Toy haulers, which can be any of the above with built-in storage for outdoor gear such as dirt bikes and ATVs.
- Teardrops, which are smaller and lighter-weight than the above, towed trailers.

The industry is booming

The Covid pandemic of 2020 made RVs and trailers wildly popular. Americans, unable or unwilling to travel in ways that put them in company with strangers, still have wanted to get out of their houses and go places. Buying or renting RVs or trailers are time-honored ways of doing that.

The RVDA reported that despite a two-month shutdown because of Covid-19, production in 2020 increased about four percent, for a yearly total of about 400,000 units. Sales have increased hugely. According to the RVIA, there was a 43.4 percent increase in sales from November 2019 to November 2020. RV Technical Institute Executive Director Curt Hemmler expects that to continue, "even as the vaccine rolls out and folks' comfort levels may return to flying again or taking a cruise." The National Association of RV Parks & Campgrounds predicted that 2021 would be a banner year; a study it conducted predicted that more than 53,000 new privately owned RV/camping sites would be constructed in 2021.

Americans are learning that RVs represent not only freedom to travel but also a good deal. A study by CBRE Hotels Advisory Group indicated that RV vacations are the most economical way to go, even with the cost of RV ownership and fuel, even if fuel prices escalate to as much as $13/gal., and even if they travel in the most expensive types of motorhomes.

Who buys RVs? The Nielsen-conducted "Go RVing Communications Planning Study" estimates that the groups most likely to be interested and to purchase RVs are "Active Family Adventurers," "Nature Lovers," and "Kid-Free Adult Adventurers," about 40 percent of all U.S. households.

My own experiences

We bought an RV when our kids were growing up and our oldest, Gabrielle, was just old enough to drive. When I say just old enough, she was all of 15 ½ years old and had a learner's permit to drive. We towed a small SUV and she would drive the RV hours and hours. She was an excellent driver, even at that age. In the first year of ownership we took that small 21 foot "Class C RV" in the middle of the winter—late December to be exact—all the way from Jefferson City, Missouri, to Key West, Florida. We were gone for ten days and had multiple night stays in Panama City, North Miami Beach, Stock Island (near Key West), Fort Lauderdale, and lastly, Orlando. It was truly the best and most complete family vacation we ever took. And most memorable.

While I was serving as executive vice president of the US Junior Chamber of Commerce, we bought a RV to put on the road for a public relations campaign entitled the "Wake Up America Tour." The rig was a 34 foot Class A, Fleetwood Bounder RV and it sported a full body wrap. Over the course of two years that RV traveled to all 48 contiguous states, some as many as seven or eight times. It hosted over 1600 town hall meetings and other events over the two-year period. The RV was the perfect "vehicle" for such an ambitious effort.

Over the past five years, I've personally owned two more RVs. One was a 34-foot Georgetown Class RV. The other, later, was a 36-foot Nexus Ghost Super C RV. It's a beast of a machine and serves as a winter, warm weather residence for first-time snow-birders. In the first two years of ownership, we put on 20,000 miles just out seeing National Parks, higher end RV Resorts and other attractive camper friendly destinations. Put me down as one who loves RVs and can see and relate to all the advantages they have to offer. RVs are an ideal extension of luxurious personal mobility.

By the numbers

The RVIA's research revealed a $114 billion overall impact on the U.S. economy. That includes nearly 600,000 jobs with $32 billion paid out in wages and $12 billion in federal, state and local taxes.

52

E-Bikes

The growth and proliferation of E-bikes

"I think e-bikes are a game changer for the industry. They allow more people to ride bikes, especially those who may not have the physical ability to ride a traditional bike."
—*Jim Sayer, Executive Director of the Adventure Cycling Association*

The growth of E-bikes in the US over the past few years has been nothing short of remarkable. E-bikes, or electric bicycles, have been around for a few decades, but only recently have they gained significant popularity in the US. This chapter explores the factors that have contributed to the growth of E-bikes in the US, the benefits and drawbacks of E-bikes, along with a forecast into the future of E-bikes.

One of the biggest factors that has contributed to the growth of E-bikes in the US has been the increased trend in urbanization. As big cities grow and more metropolitan areas build like cities and more people move into cities, there becomes a greater need for transportation that is efficient, affordable and sustainable. E-bikes offer these benefits, making them an attractive option for many city dwellers who want to get around town quickly without needing to rely on cars or public transit.

Another factor that has contributed to the growth of E-bikes is the increasing concern about the environment. As more people become aware of the impact that various transportation mobility systems have on the environment, they find themselves looking for alternative modes of transportation that are more eco-friendly. E-bikes are a popular option since they produce zero emissions and are powered by electricity, which is viewd as a much cleaner source of energy than gasoline.

The cost of E-bikes has also played a role in their growing popularity. While E-bikes can be more expensive than traditional bicycles, they are often less expensive than cars or other forms of transportation. This makes them a favored option for people who are looking for an affordable way to get around and don't have to face long commute times.

The technology behind E-bikes has improved significantly in recent years, making them more reliable, efficient and user-friendly. The batteries that power E-bikes have become more advanced, allowing them to travel longer distances per single charge. The E-bike motors have also become more powerful, allowing riders to travel faster and tackle hills and other obstacles with ease.

So, what are the benefits and drawbacks of E-bikes? Let's start with the benefits. First and foremost, E-bikes are a great way to get around quickly and easily. They allow riders to travel longer distances than traditional bicycles, without getting tired or sweaty. This makes them a great option for commuting to work or running errands around town.

E-bikes are also more eco-friendly than other forms of transportation. They produce zero emissions, which means they don't contribute to air pollution or greenhouse gas emissions, except, as do electric vehicles, through the utilization of electricity for charging. This makes them a great option for people who are concerned about the environment and want to reduce their carbon footprint.

E-bikes are a great way to stay active and healthy. While they do have a motor that provides power assistance, E-bike cyclists still need to pedal in order to move the bike forward. This means that E-bike cyclists can still get a workout while riding, which can be good for their physical and mental health.

There are, however, some drawbacks to E-bikes. One of the biggest is cost. While E-bikes can be less expensive than cars or other forms of transportation, they can still be quite expensive, especially for those that are made using the design of a high-quality bike with advanced features.

Another drawback is the range of E-bikes. While the technology behind E-bikes has improved significantly, they still can have a somewhat limited range compared to cars or other forms of transportation. This means that E-bike cyclists may need to recharge their batteries more frequently, which can prove to be inconvenient for some.

E-bikes may not be suitable for all cyclists. While they are great for commuting and running errands around town, they may not be suitable for longer rides or more challenging terrain. This means that E-bike cyclists may need to have a traditional bicycle as well as an E-bike if they want to be able to ride in all types of conditions.

New information regarding E-bikes shows a dramatic increase in personal injuries coming into doctor offices, urgent care centers and hospital emergency rooms that are resulting from accidents on these type of bikes. It makes sense to expect there to be a significant number of personal injuries.

E-bikes are heavier and faster than traditional bicycles. This can lead E-bike cyclists to ride them faster and harder than they would typically go on traditional bikes. It would stand to reason that E-bike cyclists will take greater chances at intersections, traffic mergers and other situations that could lead to accidents and injuries.

Over the past 20 or so years, I re-engaged with cycles as a sport and mobility mode for shorter excursions, as well as for training for long bicycle events, after going twenty or so years without cycling at all. With that evolution, I have experienced four cycling accidents that resulted in broken helmets, ambulance rides and hospital emergency room visits. Two of those caused broken bones (ribs, fingers, vertebrae). Luckily, none caused extended hospital stays. Three were on traditional cycles. The most recent was on an E-bike.

In that most recent incident, near Phoenix, Arizona, I asked the emergency room physician how often he sees patients brought in due to bicycle accidents. As a follow-up, I asked how many of the bicycle accident patients he sees involve cars. His response was "probably less than one in ten" involve a car. Though from a severity standpoint, those are usually the worst he sees. I have sought out data to get a better grasp of those outcomes. Meaningful data on bicycle injuries, with or without involvement with cars, is hard to locate.

I am still looking. Though I think it is safe to say that, for all the talk and seemingly justifiable concern of bicycle versus car accidents, we are much safer in our cars than on bicycles if we end up in an accident. I have never experienced an ambulance ride or an emergency room visit resulting from a car accident, let alone a car versus bicycle accident when I was occupant of a car. I have received three ambulance rides resulting from bicycle accidents while I was cycling, both with and without cars involved.

So, what does the future hold for E-bikes? It's hard to say for sure. It seems likely that they will continue to grow in popularity in the US and around the world. As more people become aware of the benefits of E-bikes, and as the technology behind them continues to improve, we can expect to see more and more people using them as a primary mode of commuting for short trips and for recreation, as well as personal fitness.

With the growth of E-bike sales and cycling growth in general, one would expect that cycling, especially bicycle commuting, would be on the rise in the US. Ironically, in a September 27 article by *Bloomberg,* statistics show that cycling to work, in fact, isn't gaining any ground in the US. Despite growth in New York and a few other big cities, commuting by bicycle is less popular nationwide than it was a decade ago. That surprises me and may surprise you too.

There are some exciting developments in the world of E-bikes that could help to drive their growth in the future. For example, there are now E-bikes designed specifically for off-road use, which could open up new possibilities for adventure and exploration.

There are also E-bikes designed specifically for people with disabilities, which could help to make cycling more accessible to everyone. And, increasingly, people are buying E-bikes that are built as cargo bikes. With these, E-bikes have increased capacity and are able to be used for shopping, dropping kids at school and other functions now provided almost exclusively by cars.

53

Emissions

The evolution of vehicle emissions – progress and regulation over the decades

"The dramatic reduction in vehicle emissions over the past 50 years is a testament to the power of innovation and regulation. We have seen a downward trend that has led to a 99 percent decrease in harmful pollutants, improving the quality of life for millions of people."
—*Union of Concerned Scientists*

Over the past five to six decades, vehicle emissions have undergone significant improvements due to a combination of regulatory measures and technological advancements. For this summary, it is my aim to explore the reasons behind this progress, including the role of the National Highway Traffic Safety Administration (NHTSA) regulations, the Clean Air Act, the California waiver allowing tougher standards, and the adoption of California emissions by certain states, both Low Emission Vehicle (LEV) and Zero Emission Vehicles (ZEV).

Additionally, the distinction between particle emissions contributing to ozone, brown cloud and asthma inducing emissions versus greenhouse gas emissions contributing to climate change will be differentiated.

Historical Context and Early Emissions Regulation:

In the mid-20th century, vehicle emissions were largely unregulated, resulting in severe environmental and public health consequences. The recognition of these adverse effects led to the implementation of initial emissions regulations in the 1960s, primarily targeting the reduction of harmful pollutants such as carbon monoxide (CO) and nitrogen oxides (NO_x). These regulations marked the beginning of a transformative journey towards cleaner transportation.

NHTSA Regulations and Technological Advancements:

The NHTSA played a crucial role in improving vehicle emissions by focusing on fuel efficiency and reducing greenhouse gases. Through the federal Corporate Average Fuel Economy (CAFE) standards, the NHTSA required automakers to increase the average fuel economy of their fleets. This resulted in the development of more efficient engines, lightweight materials, and improved aerodynamics, leading to reduced emissions of greenhouse gases such as carbon dioxide (CO_2).

The Clean Air Act and Emissions Standards:

The federal Clean Air Act, enacted in 1970 and subsequently amended, provided a comprehensive framework for addressing air pollution, including vehicle emissions. The Clean Air Act empowered the Environmental Protection Agency (EPA) to establish and enforce emissions standards for various pollutants. Over the years, the EPA has revised and tightened these standards, leading to significant reductions in harmful emissions such as particulate matter (PM), NO_x, and volatile organic compounds (VOCs).

The California Waiver and Tougher Standards:

Recognizing the need for more stringent emissions regulations, California sought and was granted a waiver under the Clean Air Act, allowing the state to set its own, more elevated, vehicle emissions standards. This waiver enabled California to implement more stringent regulations than those set by the federal government. The state's ability to establish tougher standards has been crucial in driving nationwide emissions reductions, as other states have the option to adopt either federal or California standards.

Adoption of California Emissions Standards:

Several states, known as "CARB states" (California Air Resources Board), have chosen to adopt California's stricter emissions standards. These states include New York, Massachusetts, Washington, Oregon, New Jersey, Delaware, Connecticut, Colorado and others. By aligning with California's regulations, these states demonstrated their commitment to reducing vehicle emissions and improving air quality within their jurisdictions.

Particle Emissions vs. Greenhouse Gas Emissions:

It is essential to distinguish between particle emissions and greenhouse gas emissions, as they have different environmental impacts. Particle emissions contribute to local air pollution, leading to the formation of ozone, creation of the brown cloud, and breathing/respiratory issues such as asthma. These emissions are primarily regulated through particulate matter standards.

On the other hand, greenhouse gas emissions, such as CO_2, contribute to climate change by trapping heat in the atmosphere. To combat climate change, regulations have been focused on reducing these emissions through fuel efficiency standards and incentivizing the adoption of battery electric vehicles and other zero emission cars.

Continued efforts to improve vehicle emissions will be essential in ensuring a sustainable future and progress toward an emissions free transportation sector.

Aside from getting newer and much cleaner cars on the road, year-in and year-out, there is also the issue of the older, less clean, often high emitting vehicles that were added to the roads over the past several decades. New cars are so much cleaner; the old ones they replace can be 100 times harder on the environment than the new ones coming into service today. A national or even international effort is needed to target the old high emitters and take them off the road so they can be replaced with newer cleaner cars.

New car dealers in Colorado took the bold step in 2007 to fund and launch the startup of a non-profit, tax-exempt foundation charged with recycling old high polluting cars and replacing them with the next generation new environmentally friendly fleet. It's appropriately named the Clear the Air Foundation. Between 2007 and today, that foundation has recycled over 7,300 high emitters and with the proceeds funded scholarships for auto tech students at levels of $2500 to $5000 and has contributed over $300,000 so far toward the effort of getting the next generation of techs into service bays. No other dealer group nor other foundation has stepped up in this way. Yet this program could, with the right guidance, grow to become a nationwide effort.

54

Road safety

The evolution of automotive safety

Advancements in vehicle safety technology have drastically reduced traffic fatalities versus vehicle miles traveled (VMT), saving countless lives and making roads safer for everyone, though recent trends of fatalities by cyclists and pedestrians threatens to reverse the downward fatality trend of traffic safety.

The automotive industry has witnessed remarkable advancements in vehicle safety over the past few decades. Today, new cars are equipped with sophisticated safety features and technologies designed to minimize the risk of accidents and protect occupants in the event of a collision. In this chapter we explore the significant strides made in automotive safety by comparing the annual traffic fatality rates of the early 1970s to the present day. Through examination of statistical data and analyzing the advancements in vehicle safety, it becomes evident how much safer new cars are today.

- State of motor vehicle safety in the 1970s: In the early '70s, automotive safety was far from being a forefront concern. Cars lacked essential safety features and the overall design prioritized aesthetics and performance over occupant protection. Seatbelts were often optional and airbags, having only just been perfected as safety devices, were virtually nonexistent. Additionally, vehicle structures lacked the necessary reinforcement to withstand high-impact collisions, resulting in passengers being highly vulnerable to severe injuries in medium to severe crashes. Starting in 1972, the federal government required seat belts in all motor vehicles. That action drove a sharp reduction in traffic fatalities over the next 40+ years.

- Advancements in crashworthiness: The concept of crashworthiness, or a vehicle's ability to protect occupants during an accident, has undergone significant improvements since the early 1970s. The introduction of mandatory safety features such as seat belts, airbags and reinforced structures has played a significantly pivotal role in reducing fatalities. As an example, in 1973, only 11% of vehicles were equipped with seat belts, while today, nearly all new cars are equipped with these life-saving features.

- The positive impact of seat belt usage: One of the most significant factors contributing to the reduction in traffic fatalities is the increased usage of seat belts. In the early 1970s, seat belt usage was low due to lack of awareness and mandatory laws. However, as public awareness campaigns and legislation promoting seat belt usage gained momentum, the number of lives saved increased exponentially. According to the NHTSA, seat belts saved an estimated 14,955 lives in the year 2017 alone.

- Emergence of airbag technology: The introduction of airbags in the early 1980s further revolutionized vehicle safety. Initially, airbags were installed only in luxury vehicles but gradually became a standard feature in most cars. Airbags significantly reduce the risk of fatal injuries for in-vehicle passengers by cushioning the impact during a collision. Studies have shown that frontal airbags alone can reduce the risk of driver fatalities by approximately 29%. Some vehicles today come equipped with as many as 14 airbags.

- Advancements in vehicle structural integrity: Vehicle structures have also undergone substantial improvements to enhance crashworthiness. The incorporation of high-strength materials, such as advanced steel alloys and aluminum, has enabled automakers to design vehicles with greater structural integrity. These advancements have resulted in improved energy absorption and dispersion during a crash, reducing the likelihood of severe injuries or fatalities.

- Active safety technologies: Besides passive safety features, new cars are equipped with a range of active safety technologies designed to prevent accidents altogether. Antilock Braking Systems (ABS), Electronic Stability Control (ESC), Forward Collision Warning (FCW) and Lane Departure Warning (LDW) are just a few examples of such technologies. These systems help drivers avoid potential collisions and mitigate the severity of accidents, further reducing the risk of fatalities.

- The 40-year statistical decline in traffic fatalities was followed soon by a sharp upward tick. To better understand the extent of safety improvements, an analysis of the statistical decline in traffic fatalities over the years is important. In the early 1970s, the United States recorded an average of approximately 54,000 traffic fatalities per year. By 2014, this number had decreased to around 32,000 fatalities annually. This represents a significant reduction of more than a third of traffic fatalities over the span of nearly five decades. However, by 2020, the annual traffic fatalities count had increased back up to about 48,000, a significant increase over 2014, largely due to increased fatalities involving pedestrians and cyclists.

- Trends over the past century based on fatalities versus population: The population motor-vehicle death rate reached its peak in 1937 with 30.8 deaths per 100,000 population. The current rate, based on 2021 data, is 14.3 per 100,000, representing a 54 percent improvement. In 1913, 33.38 people died for every 10,000 vehicles on the road. In 2021, the death rate was 1.66 per 10,000 vehicles, a dramatic and impressive 95 percent improvement.

The bottom line is this: Cars continue to become safer and safer for occupants, even as traffic casualty counts rise due to an increase in fatalities among cyclists and pedestrians. As more and more people populate ever more crowded streets and more pedestrians and cyclists are injected into the traffic-way puzzle, new onboard vehicle technologies and transportation system design will play an ever-growing and more critical role in effective traffic fatality reduction efforts.

Advancements in automotive safety have transformed the land-scape of vehicle design, resulting in significantly safer cars on the roads today. Motor vehicles will continue to become safer for their occupants over time, and with better onboard safety and detection technologies, will become significantly safer for cyclists and pedestrians.

55

Quality reigns

Evolution of new-car quality – paving the way for extending vehicle longevity

"Over the past few decades, advancements in technology and engineering have led to a significant increase in the quality of new cars. This has resulted in improved vehicle longevity, durability, lifespan and reliability, providing consumers with vehicles that not only last longer but also require fewer repairs and maintenance, ultimately saving a lot of time and money."
—Lee Payne, Planet Honda, Planet Hyundai, Genesis of Golden

The automotive industry has witnessed remarkable advancements in vehicle quality over the past several decades. New cars have experienced significant improvements in terms of durability, reliability and overall quality. These advancements have played a pivotal role in increasing the longevity of vehicles and, subsequently, the average age of the overall fleet on the road in the US. This chapter covers the evolution of new car quality over the past 40, 60 and even 80 years, while analyzing the correlation between improved quality and increased vehicle longevity.

Design

From a design standpoint, new cars have seen substantial enhancements in quality. The implementation of advanced manufacturing techniques, improved material content and stricter quality control measures have been instrumental in this transformation. Automakers have focused on addressing common issues such as engine reliability, rust-prevention and overall build quality. The integration of computer-aided design, precision-engineering and robotic manufacturing processes significantly reduced human error and improved overall product consistency.

The advancements in the quality of new cars, from a design standpoint, have been even more pronounced. This period witnessed the advent of numerous technological breakthroughs, including the introduction of electronic fuel injection, power steering, disc brakes and electronic ignition systems. These innovations not only improved performance but also enhanced the reliability and longevity of vehicles. Additionally, advancements in safety features such as seat belts, airbags and anti-lock braking systems (ABS) have contributed to the overall quality and value of new cars over older, less-improved models.

Engineering

From an engineering standpoint, the improvements in automotive quality have been truly transformative. The introduction of mass production techniques by Henry Ford in the early 20th century revolutionized the industry, making vehicles more affordable and accessible to the general public. As the industry evolved, automakers focused on addressing quality issues such as frequent breakdowns, poor fuel efficiency and susceptibility to rust. The development of more robust engines, corrosion-resistant materials and improved manufacturing processes significantly enhanced the quality and longevity of new cars.

Significant improvements in new-car quality have had a profound impact on vehicle longevity. In the past, it was not uncommon for cars to become unreliable and reach the end of their lifespan within a decade. However, with the advancements in quality, modern vehicles are built to last longer and require fewer repairs. The improved durability of engines, transmissions and other critical components helps ensure that new cars can withstand higher mileage and a wider range of driving conditions. Consequently, vehicle owners can now expect their cars to remain functional for many more years, leading to an increase in the average age of vehicles on the road.

Quality

From an overall quality standpoint, cars have improved remarkably over the years. The average age of the overall fleet of vehicles on United States roads has steadily increased over the past six to eight decades. This can be attributed to the improvements in new-car quality discussed earlier. According to data from the Bureau of Transportation Statistics, the average age of vehicles on the road in the United States has risen from 6.1 years in 1970 to more than double that, 12.6 years, in 2022. This upward trend can be primarily attributed to the improved longevity of vehicles resulting from enhanced quality standards.

While improved quality has been a significant factor in increasing vehicle longevity, other factors also contribute to this trend. Regular maintenance, technological advancements and changing consumer behaviors have all played a role. The availability of better lubricants, more efficient cooling systems and improved maintenance practices have helped extend the lifespan of vehicles. Additionally, the shift toward a more service-oriented economy, where consumers prioritize maintaining and repairing their existing vehicles rather than buying new ones, has also contributed to the increased average age of the fleet.

Real-life verification

In everyday and practical terms, this story I heard firsthand during business travel to Gunnison, Colorado, some 15 years ago, demonstrates the significant, even monumental improvement in new-car quality better than any I know. While speaking to the Rotary Club of Gunnison, I addressed the issue in automotive culture of improved quality of cars and increasing age of fleet. In the audience was a long-time Gunnison resident, who was the former owner of a local General Motors dealership, with most GM brands included in their franchise agreements, including Chevrolet, Buick, Pontiac, Oldsmobile and Cadillac. His family was in the new-car dealer business in Gunnison for several decades through the 1940's, '50's, '60's and '70's.

After I told the story of how new-car quality had led to greater longevity of all aspects of the automotive fleet, this former dealer spoke up from the audience. I had clearly hit a chord. The former dealer referenced how in the '50s, '60s and even early 70s, new cars would come off the delivery truck when arriving at the dealership and would have go through two or three days of quality control inspections, pre-delivery repairs and changes before the cars could be handed off to their customers. He said on average each car would have 30 to 40, sometimes even 50, specific repairs or changes needed before they could be delivered to their ultimate owners. By comparison, it is rare today for new cars to need any repairs or modifications after they arrive from the factory. We have come so far. New cars produced today are ready to roll, often with no repairs needed for 100,000 miles or more.

Warranty claims tell the story

One other meaningful trend further demonstrates the monumental improvement in new-car quality over that past few decades. This one is anecdotal yet so strong on how much quality has improved. A Colorado-based Chrysler Jeep Dodge Ram dealer who opened his dealership in the mid-90s and sold it in 2014, from an ongoing process beginning on Day One, kept track of the percentage of his service bay business that was for warranty claims versus customer pay. He kept the records year-in and year-out. When he started this tracking in the mid-90s, over 60 percent of the store's service bay repair business was reimbursed by the factory as warranty pay, leaving less than 40 percent as customer pay. Each year the warranty pay percentage declined versus customer pay revenue. By the time he sold the dealership the warranty pay percentage had dropped to less than 15 percent of total service bay business. That is a meaningful and impressive decline. And it is one that further demonstrates the dramatic improvement in the quality of new cars.

That improvement in quality has been across brands, makes and models. During my years in the business, I was often asked which cars a friend or relative should avoid. I had to respond with the facts. The facts are that there is not a bad car in today's new car market. A bad car would not survive in today's marketplace. Social media and customer reviews would literally ruin the business of anyone building or selling cars of low quality in today's market.

Articulated in expected vehicle miles, new cars of 30 to 50 years ago were considered old and not worth much in dollar value once they had been driven 80,000 to 100,000 miles. At that point they were of little value to trade to a dealer or to pass down (or up) to a friend or relative. At 100,000 miles, a 1970 model car was ready for the salvage yard. By comparison, a car built in the last 15 to 20 years and with 100,000 miles on it is truly barely broken in. A new car today may run 250,000 to 300,000 miles before it needs any kind of repair or auto tech. At 100,000 miles, a car today is certainly not ready for a salvage yard and still has many quality-of-life miles ahead.

Bottom line: New-car quality has improved exponentially. It's been gradual, not necessarily overnight. Though it has happened. That significant improvement in quality has changed how we buy cars, how long we keep them, what we need to do to them while we own and operate them, and how long they ultimately last for service on the road. The change has been meaningful, even monumental. And it has been very good for car buyers, owners and drivers.

Have we passed peak longevity with EVs?

Have we achieved peak automotive quality and longevity improvements, causing a new trend of reversing growth of vehicle age curve? New cost reports for some BEV models, such as Teslas are jaw-droppingly high. Tesla repairs are more expensive than average, in part, because the electric vehicle battery is a very expensive component.

One recent report shows the average cost of repairing an electric powered Tesla is $5,552, an amount of $1,347 more than comparable internal combustion vehicles and $1,078 more than competitive, non-Tesla, electric vehicles. Specific auto insurance companies have announced plans to stop insuring electric vehicles.

The increasing complexity of vehicles is making repairs more challenging and potentially more expensive. This can encourage some vehicle owners to keep older vehicles on the road longer than they have in the past.

I believe cars with sophisticated computer-driven features are not going to be affordable after 10-15 years of vehicle ownership. It may be less expensive to take them out of service and recycle the salvage materials to scrap them then to fix them, similar to modern-day appliances.

Based on current economic trends and new technologies, I believe we are about to pass peak auto longevity.

56

Electrification

The electrification of the auto industry – a bumpy journey toward more sustainable transportation

"The future of passenger vehicle powertrains is electric; the transformation is ongoing."
—*McKinsey and Company*

The electrification of the auto industry has been a long and winding road, marked by failed attempts, incremental progress and recent dramatic increases in both supply and demand for battery electric vehicles (BEVs).

This chapter aims to delve into the historical context of battery electric cars, examining the early days of their development, the failed attempts to electrify the industry in the past and the recent surge in EV adoption. Additionally, it will explore the governmental efforts, both in the United States and globally, to require and/or incentivize (or both) the use of BEVs, with a particular focus on California's regulations that have been adopted by several other states.

Early days: The birth of electric cars in the auto industry

The concept of BEVs dates all the way back to the earliest days of the automobile industry. In the late 19th century, electric cars gained popularity due to their simplicity, ease of use and lack of noise and emissions. Early electric car pioneers such as Thomas Davenport and Thomas Parker were instrumental in developing the first practical electric cars, introducing innovations such as rechargeable batteries and electric motors. The first Ford Model Ts were built as BEVs.

As I referenced in the chapter covering Henry Ford and the start of Ford Motor Company, one of Ford's closest and most influential friends was none other than Thomas Edison. Also, instrumental to this decision-making was that Henry Ford's wife, Claire, owned an early all-electric car. Both Claire Ford and Thomas Edison encouraged Henry Ford to build the first Model Ts as a BEV. Ford's reluctance then, is the same pushback we hear today, some 120 years later … they have limited range, they take too long to charge, charging infrastructure is not adequate and they cost too much.

The first auto shows in the United States were started in New York City in 1900, Chicago in 1901 and Denver in 1902. By the way, the Detroit Auto Show and the Los Angeles Auto Show both started in 1907. In the first Denver Auto Show in 1902, there were only 29 cars on display. Of those, about half were steam-powered, and the other half were battery-electric-powered. None were gas-powered.

Failed attempts in recent years

Despite the promise shown by early BEVs, the industry faced numerous challenges that impeded the quickest growth and most widespread adoption of BEVs. The 1970s oil crisis sparked renewed interest in electric vehicles due to concerns over energy security and rising fuel prices. However, limited battery technology, range anxiety and the lack of charging infrastructure continued to hinder progress toward a shift to electric vehicles. In the 1980s, some automakers attempted to introduce BEVs, but these efforts were unsuccessful.

The 1990s witnessed a resurgence of interest in electric vehicles, driven by concerns over air pollution and climate change. Automakers, such as General Motors with the EV1 and Toyota with the RAV4 EV, made new attempts to popularize electric cars. However, despite improvements in technology, these initiatives were short-lived due to limited consumer demand and the lack of supportive policies and infrastructure.

General Motors' EV1 and other early electric models faced obstacles such as high costs, limited range, and the lack of consumer awareness. As a result, these programs were discontinued, and the auto industry shifted its focus back to further refinements and improvement to internal combustion engines.

Hybrid-powered cars, which derived energy from an internal combustion engine as well as a battery were popularized almost solely by Toyota Motor Corporation. The Toyota Prius launched in 2000 and has become the quintessential pure hybrid vehicle.

Pure hybrid-powered cars expanded to Plug-In Hybrid Vehicles (PHEVs), wherein a hybrid car can plug in to the grid at night and operate the next morning on battery power, at least for a short distance, generally 20 to 30 miles. Again, Toyota has been the leader in this technology, though the technology has recently gained traction with Stellantis, Hyundai, Kia, BMW, Volvo, Honda and more. Both Ford and General Motors have had successful hybrid cars, though those have gone away for the most part in favor of full BEVs. Early in 2024, Toyota has doubled down on hybrid technology vehicles and both General Motors and Ford and made moves to invest deeper in hybrid technology due to their strong marketplace popularity.

Current surge: supply and demand for battery-electric vehicles (BEVs)

The early 21st century marked a turning point for battery-electric vehicles, with advancements in battery technology, cost reductions and increasing environmental consciousness among consumers.

Tesla's BEVs, first introduced in 2008, demonstrated the potential of BEVs by offering impressive performance and a reasonable range. This paved the way for other automakers to invest in developing their own BEVs. The past decade has witnessed a rapid increase in both the supply and demand for battery-powered vehicles as well as hybrid and PHEVs. Automakers such as Tesla, Nissan, General Motors, Ford, Hyundai, Kia, Volvo, VW and BMW have introduced affordable electric models with improved range and performance. Toyota has been a leader in hybrid and PHEVs and BMW has been a leader in PHEVs.

Further, the rise of shared mobility services and the increasing availability of charging infrastructure, along with the longer range of electric vehicles, have addressed some of the concerns surrounding electric vehicle adoption. Still today, the greatest reluctance of many new car buyers when considering full battery-electric vehicles are: 1.) range, 2.) higher cost, 3.) charging availability, 4.) amount of time it takes to charge.

Governmental efforts: mandates plus incentives

Recognizing the urgency of transitioning to sustainable transportation, governments worldwide have implemented policies and incentives to promote BEVs. The push to EVs has been acute and effective in China and across Europe, especially in the Scandinavian countries of Norway, Finland and Sweden.

In the United States, the federal government has provided tax incentives for electric vehicle purchases, while states such as California have taken the lead in setting extremely ambitious targets for zero-emission vehicle sales. California's regulations have had a significant impact on the auto industry, as several other states have adopted similar standards.

California's Zero Emission Vehicle (ZEV) program, established in 1990, requires automakers to produce a certain percentage of electric and other zero-emission vehicles. This has spurred innovation and investment in the sector, leading to a broader range of electric vehicle options for consumers.

California was only able to create its own standard due to a waiver allowed by the EPA dating back to the 1970s. That original waiver has been updated several times since. About 15 other states have also adopted either California's Low Emission Vehicle (LEV) or ZEV, or both. Having two different standards across the country has caused a lot of consternation across the political spectrum.

Automotive advocacy groups have largely opposed having a separate vehicle emissions standard established at the state level and exported to multiple other states that is separate from a nation-wide federal standard. At least one state level automotive advocacy group has moved from a long-time position opposing the California standard, to one of neutrality. That group even considered and voted on endorsing the California standard. The group ended up staying neutral though still stands as the only one almost on all sides of the California emissions regulation issue over time.

The electrification of the auto industry has come a long way from the early days of battery-electric vehicles that saw a lot of political and public pushback to the current surge of both increased supply and demand for electrified models, including hybrid, PHEVs and BEVs. Still, the stated goals of the US government along with some of the states under the California emission standards will be very difficult, if not impossible, to achieve.

For example, in January 2019, new Governor Jared Polis's first executive order called on Colorado to adopt the California ZEV standard and to put 940,000 electric vehicles (BEVs and PHEVs count) on the road by 2030. Almost five (5) years into that very lofty goal, the state had only been able to get about 100,000 new electric vehicles of the 940,000 on the road. Even being the Number Five (#5) state in the nation for EV adoption, the stated goal will have to look like a hockey stick on a chart to be possible.

Personally, I like electric vehicles a lot. They are quiet, perform extraordinarily well and come with lots of onboard technology that is generally optional on gas-powered cars. In the past three years, I have owned or operated five different electric vehicles, whether as company cars or personal, I've owned four (BEVs) and one PHEV. I can honestly say these have been some of the best cars I've experienced in my lifetime.

The long-held promise of price parity for battery-electric versus gas-powered vehicles has yet to materialize. Ultimately, aside from range, charging time and access to chargers, I think the real pushback on EVs is that of cost to purchase. Even with lower operating costs, the barrier to entry is cost to initially access. Eventually we need to get to price parity between electric vehicles and ICE powered cars.

Counterview to the electric car advocacy for greater balance:

Electric vehicles (EVs) are often touted as a more environmentally friendly alternative to traditional gasoline-powered cars, but it's essential to acknowledge that they are not entirely devoid of environmental impact. Here are some key points as to how EVs can be worse for the environment in certain contexts:

Battery production: The manufacturing of BEV and PHEV batteries, particularly lithium-ion batteries, can have a significant environmental footprint. The extraction of raw materials, such as lithium, cobalt and nickel can lead to habitat destruction, water pollution and other environmental issues. Additionally, the energy-intensive battery production process can result in substantial greenhouse gas emissions.

Electricity generation: The environmental impact of BEVs largely depends on the source of the electricity used to charge them. In regions where electricity is primarily generated from coal or other fossil fuels, BEVs may indirectly contribute to carbon emissions. However, as the electricity grid becomes cleaner, less dependent on coal and natural gas to generate electricity and more reliant on renewable energy sources, this concern diminishes.

Battery disposal and recycling: The end-of-life management of BEV batteries can be problematic. Proper disposal and recycling are crucial to minimize environmental harm, but these processes are still under development and mismanagement could lead to pollution and waste issues.

Manufacturing processes: The manufacturing of BEVs can involve energy-intensive processes. If the electricity used for manufacturing is primarily from fossil fuels, it can increase the carbon footprint of the vehicles.

Indirect emissions: The overall environmental impact of BEVs may also be influenced by factors such as how they are used and charged. If they lead to increased electricity demand and strain on the grid, it could indirectly contribute to emissions, especially in areas with an insufficient renewable energy infrastructure.

It's important to recognize that the environmental impact of EVs is highly dependent on regional and technological factors and based on how the electricity is produced that is used to power the EVs.

There are huge political divisions in public opinion regarding EVs. This is a battleground largely played out regionally, in a Blue City versus Red State partisan political divide over transportation and mobility. Each side has valid arguments, for sure. However, the divide is costly for automakers, the auto industry in general and ultimately for consumers.

The fact that new cars today are much easier on the environment and air quality—99 percent cleaner on brown cloud, ozone enhancing, asthma induced emissions—cannot be argued. It's the science and the fact. Yet, it can also be argued, that without government mandates on tailpipe emissions the industry would have never evolved to that point. That is also a very solid fact.

The policy question going forward is, how far and how fast, and at what cost can and should governments further tackle tail-pipe emissions and do so not only on cars, but on trucks, trains, ships, planes and basically every mode of mobility available to mankind. That is a costly and important question. And not one we can resolve in this chapter and book. Just know that vehicle electrification is the $64 trillion question for policymakers going forward. The public and all interest groups should weigh in and be heard. And this is the debate of transportation for the next few years and even decades.

57

Franchised vs Direct

Franchised dealer model for cars vs direct sales

"The franchised dealer distribution model allows for personalized and hands-on approach to sales and service, providing customers with a unique and tailored experiences that cannot be replicated through direct sales model."
—Anonymous salesperson at a factory direct store

Some of the earliest franchised dealers, who operated with what is called a sales and service agreement, date to the latest years in the 1800's, around 1898. Oldsmobile, as covered in Part One of this book, had franchised dealers in 1898. Henry Ford's car company, Ford Motor Company, had independent franchised dealerships as early as 1908. At least one of those is still in business as a dealership in Montrose, Colorado today, Flower Motor Company, though with other franchised brands today. Another one I know, started as a franchised Ford dealer in the Denver market in 1913 and still operates today as a fourth generation Ford dealer, O'Meara Ford.

The franchised dealership system is a critical aspect of the automotive industry, providing consumers with a wide range of benefits that ensure they have access to dealerships close to home, the ability to negotiate the best price for the sale and service of motor vehicles and the convenience of keeping services near by.

Here, we delve into the importance of the franchised dealership system in providing these essential benefits to consumers. One of the primary benefits of the franchised dealership system is the accessibility it provides to consumers. By having a network of dealerships spread across various locations, consumers can easily find a dealership close to their home, making it convenient to purchase and service their vehicles. This accessibility is crucial, as it ensures that consumers have easy access to the resources they need to buy and maintain their vehicles.

The franchised dealership system allows for competition among dealerships, bothe interbrand (same brand dealerships) and intrabrand (competing brands in market) which can ultimately benefit consumers. With multiple dealerships in a given area, consumers have the opportunity to shop around and negotiate the best transactional price for their vehicle or service.

This competitive environment encourages dealerships to offer competitive pricing and incentives to attract customers, ultimately leading to better deals and savings for consumers. In addition to providing accessibility and competitive pricing, the franchised dealership system also ensures that consumers have access to reliable and quality service close to home.

With a network of franchised dealerships in various locations, consumers can easily find a dealership that offers the services they need, such as routine maintenance, repairs and/or warranty work. This accessibility to service is essential, as it allows consumers to keep their vehicles in good condition without having to travel long distances or incur additional expenses.

The franchised dealership system also provides consumers with the assurance of receiving quality service from trained and certified professionals. Dealerships within the franchise system are required to adhere to strict standards and guidelines set forth by the manufacturer, ensuring that consumers receive high-quality service from knowledgeable technicians. This assurance of quality service is crucial, as it provides consumers with reliable peace of mind knowing that their vehicles are in good hands.

Another important aspect of the franchised dealership system is the availability of genuine parts and accessories. Dealerships within the franchise system have access to a wide range of genuine parts and accessories from their affiliated automaker that are specifically designed for the vehicles they sell. This ensures that consumers can easily find the parts and accessories they need to maintain or customize their vehicles, without having to worry about the quality or compatibility of the products.

In addition to providing accessibility, competitive pricing, and quality service, the franchised dealership system also plays a significant role in supporting the local economy. By having a network of dealerships in various locations, the franchised dealership system creates job opportunities and contributes to the economic growth of the communities in which they operate. This economic impact is essential, as it helps to stimulate local economies and support small businesses within the community.

The counter to the independently franchised dealer network is that of factory direct sales outlets for sales, delivery and service. That system, often known today as the "Tesla model" is certainly the counter to the independent franchised dealer model. The challenge to the manufacturer is having to carry all of the costs for a huge distribution network, nationwide or even worldwide. That requires a lot of extra, otherwise unneeded, capital. Sometimes consumers will say they favor the direct sales model as there is no negotiation on price—all cars are sold at the manufacturer's suggested retail price (MSRP.) According to some studies, over 90 percent of all new cars are sold under MSRP. If so, that is a huge reason for those consumers who support the franchised dealer distribution system—they can save a lot of money.

It could be stated that the franchised dealer network system of distribution creates a win, win, win. It is a win for the customer, due to lower price, closer to home services and knowing the person or people they are dealing with. It's a win for automakers, as they do not need to invest a multi-trillion-dollar investment, collectively, in dealership showrooms, service facilities and localized infrastructure around the world. And it's a win for the communities franchised dealers serve. Those dealerships invest in their communities, provide charitable and community service and become part of the local fabric and infrastructure of their home base neighborhoods.

The franchised dealership system is of utmost importance for consumers, as it provides them with access to dealerships in their neighborhood, the ability to negotiate the best price for the sale and service of motor vehicles, and the convenience of keeping services nearby. This system ensures that consumers have easy access to the resources they need to buy and maintain their vehicles, while also supporting the local economy. As such, the franchised dealership system plays a critical role in ensuring that consumers have a positive and fulfilling experience when purchasing and servicing their vehicles.

58

—

The top influencers today in personal mobility

The automotive, transportation and mobility industries have undergone significant transformation in recent years, driven by technological advancements, environmental concerns and changing consumer preferences. This list highlights those impressive individuals that I view as among the most influential voices today in the fields of automotive, transportation and mobility. These various perspectives and diverse voices encompass a wide range of expertise and interests that include cars, trucks, recreational vehicles (RVs), bicycles, motorcycles, private planes and even vertical takeoff and landing vehicles (VTOLs), as they all contribute to the broader transportation/mobility ecosystem and, ultimately, personal mobility today and in the future.

Organizations/Companies (5)
American Automobile Association (AAA)
Automotive News
Aviation News
Automotive Trade Association Executives (ATAE)
National Automobile Dealers Association (NADA)

Associations/Chairs/Leaders – Associations Drive America
Mike Stanton, CEO of National Automobile Dealers Association (NADA)
Gary Gilchrist, Chair of National Automobile Dealers Association (NADA)
Jennifer Colman, President of Automotive Trade Association Executives (ATAE)
John Devlin, President/CEO of Pennsylvania Automotive Association (PAA)
Jim Appleton, Jr., President of New Jersey Coalition of Automotive Retailers (NJCAR)
Jim Addis, Idaho Automobile Dealers Association (IADA)
Rod Albert, President of Detroit Automobile Dealers Association (DADA)
Jared Allen, NADA Communications (NADA)

Marsha Allen, Executive Vice President of Wyoming Automobile Dealers
Association (WADA)

Bruce Anderson, President of Iowa Automobile Dealers Association (IADA)

Mark Baker, President of Aircraft Owners and Pilots Association (AOPA)

Dan Bennett, President of New Hampshire Auto Dealers Association (NHADA)

Craig Bickmore, President of New Car Dealers Association of Utah (NCDU)

Gene Boehm, CEO of AAA

Anthony Brownlee, Chairman of American International Automobile Dealers
Association (AIADA)

Tom Brown, President of Maine Automobile Dealers Association (MADA)

Terry Burns, President of the Michigan Automobile Dealers Association (MADA)

David Cardella, CEO of Colorado Independent Automobile Dealers Association
(CIADA)

Evelyn Cardenas, President/CEO of Central Florida Automobile Dealers Association
(CFADA)

Jonathan Collegio, Executive Vice President of NADA Communications (NADA)

Pamela Crane, San Antonio Automobile Dealers Association (SAADA)

Tom Dart, President of Alabama Automobile Dealers Association (AADA)

Zach Doran, President of Ohio Automobile Dealers Association (OADA)

C. Brent Franks, President of North Texas Automobile Dealers Association
(NTADA)

Sims Floyd, President of South Carolina Automobile Dealers Association (SCADA)

Greg Fulton, President of Colorado Motor Carriers Association (CMCA)

Jill Goldfine, Automotive Trade Association Executives (ATAE)

Matthew Groves, CEO of Colorado Automobile Dealers Association (CADA)

Don Hall, President of Virginia Automobile Dealers Association (VADA)

Peter Hodges, President of Oklahoma Automobile Dealers Association (OADA)

Phil Ingrassia, President of Recreation Vehicle Dealers Association (RVDA)

Greg Kirkpatrick, President of Arkansas Automobile Dealers Association (AADA)

Lea Kirschner, President of Georgia Automobile Dealers Association (GADA)

Pete Kitzmiller, President of Maryland Automobile Dealers Association (MADA)

Bruce Knudsen, President of Montana Automobile Dealers Association (MADA)

Andrew Koblenz, Executive Vice President of National Automobile Dealers
Association (NADA)

Scott Lambert, President of Minnesota Automobile Dealers Association (MADA)

Matthew Larsgaard, President/CEO, Automobile Dealers Association of North
Dakota (ADAND)

Vicki Fabre, President of Washington State Automobile Dealers Association
(WSADA)

Cody Lusk, President of American International Automobile Dealers Association
(AIADA)

Brian Maas, President of California New Car Dealers Association (CNCDA)

Andrew Mackay, President of Nevada Franchised Automobile Dealers Association
(NFADA)

Mike Martin, CEO of National Independent Automobile Dealers Association
(NIADA)

Marty Milstead, President of Mississippi Automobile Dealers Association (MADA)

Skyler McKinley, Chief Spokesperson of AAA – The Auto group Club

Dan McNeeley, President of Kansas Automobile Dealers Association (KADA)

Paul Metrey, NADA Senior Vice President of Legal and Policy (NADA)

Jennifer Morand, General Manager of Chicago Auto Show and Chair of ASNA

Marty Murphy, President of Indiana Automobile Dealers Association (IADA)

John O'Donnell, President and CEO of Washington Area New Automobile
     Dealers Association (WANDA)

Robert O'Koniewski, President of Massachusetts State Automobile Dealers
     Association (MSADA)

Ken Ortiz, Executive Director of New Mexico Automobile Dealers Association
     (NMADA)

T. John Policastro, President of North Carolina Automobile Dealers Association
     (NCADA)

Geoff Pohanka, Immediate Past Chair of National Automobile Dealers Association
     (NADA)

Greg Remensperger, President of Oregon Automobile Dealers Association (OADA)

David Regan, Executive Vice President of NADA

Ivette Rivera, Senior Vice President of National Automobile Dealers Association,
     (NADA)

Mark Scheinberg, President of Greater New York Automobile Dealers Association
     (GNYADA)

William (Bill) Sepic, President of Wisconsin Automobile and Truck Dealers
     Association (WATDA)

John Sackrison, President of Orange County Automobile Dealers Association
     (OCADA)

David Sloan, CEO of Chicago Automotive Trade Association (CATA)

Bob Smith, President Greater Los Angeles Automobile Dealers Association
     (GLADA)

Ted Smith, President of Florida Automobile Dealers Association (FADA)

Doug Smith, President of Missouri Automobile Dealers Association (MADA)

Bobbi Sparrow, President of Arizona Automobile Dealers Association (AADA)

Robert Vancavage, President of New York State Automobile Dealers Association
     (NYSADA)

Lou Vitantonio, President of Greater Cleveland Automobile Dealers Association
     (GCADA)

Dennis Whitehurst, President of Texas Automobile Dealers Association (TADA)

Transportation, Automotive and Mobility Executives – Advocating Freedom to Drive

Melissa Kuipers, Partner/Atty Brownstein Hyatt Farber and Schreck, rising superstar

Eric Anderson, Partner at SE2

Tim Cook, CEO of Apple

Paul Daly, Founder of Automotive State of the Union (ASOTU)

Curtis DeGroote, CEO of Marketpoint

Veronica Dunford, CEO of Women in Automotive

David Kain, Senior Advisor NCM Digital Performance Solutions
Dara Khosnowshahi, CEO of Uber
Sundar Pichait, CEO of Google
Peter Johns, Xcel Energy
Mike Marshall, Success Amplified Business Coaching
Gary May, Founder/President of Interactive Marketing and Consulting Services
John Murphy, Head of North American Automotive Research at Bank of America
Roman Mica, Nathan Adlen and Andre Smirnoff, Leadership team at The Fast Lane
    Cars and Trucks
Kyle Mountsier, Partner/COO of Automotive State of the Union (ASOTU)
Eric Miltsch, VP and SEO od Dealer Teamwork
John Murphy, Head of North American Automotive Research at Bank of America
George Nenni, Generations Digital
Greg Noonan, Customer Retention and Loyalty Consultant
Brian Pasch, Founder of Pasch Group
Glenn Pasch, Partner in Pasch Group
April Rain, Automotive Marketing Executive
Stephen Rowling, President of Cox Automotive
Jennifer Sanford, Automotive Marketing Consultant
Jonathan Smoke, Chief Economist at Cox Automotive
Todd Smith, Qore AI
Troy Spring, CEO and Founder of Dealer World
Alex Taylor, Chairman of Cox Automotive
Sean Ugrin, Xcel Energy
Alex Vetter, CEO of Cars Commerce
Charlie Vogelheim, automotive and VTOL (flying cars) expert and enthusiast
Joe Webb, Founder of Dealer Knows Consulting
Jim Ziegler, CSP, HSG, OG, President of Zieger Supersystems

Governmental Leaders – Understanding Freedom of Mobility
Maria Cantwell, Chair of US Senate Transportation Committee
Jack Danielson, Director of USDOT National Highway Transportation Safety
    Administration
Sam Graves, Chair of US House Transportation Committee

Aviation Executives – Freedom to Fly
Lisa Atherton, CEO of Bell Helicopter
JoeBen Bevirt, Founder of Joby Aviation
Johann Bordais, CEO Embraer Eve Mobility
Kyle Clark, Founder and CEO, Beta Technologies
Stephen D'haene, CEO of Jetson One
Adam Goldstein, CEO of Archer Aviation
Dr. Bryan Yutko, CEO of Wisk Aero

Automotive Executives – Freedom to Drive
Mary Barra, CEO of General Motors
John Bozella, CEO of Alliance of Automotive Innovation

Jim Farley, CEO of Ford Motor Company
Bill Ford Jr., Chairman of Ford Motor Company
Jack Hollis, President of Toyota North America
Pablo Di Si, President/CEO of Volkswagen Group of America
Noriya Kaihara, CEO of Honda of Americas
John Krafcik, Board Member of Rivian, Founding CEO of Waymo
Sebastian Mackensen, CEO of BMW of North America
Dimitris Psillakis, CEO of Mercedes-Benz USA
Elon Musk, Chairman of Tesla and founder of SpaceX
Mark Reuss, President of General Motors of North America
Kaji Sato, CEO of Toyota
Carlos Tavaras, CEO of Stellantis
Chris Urman, CEO of Aurora (autonomous drive technology)

Automotive, Transportation and Mobility Media – Covering all ways of getting around

Jason Stein, Host/Managing Director of Presidio Full Throttle and host of *Cars and Culture*, Sirius XM
Jay Leno, founder *Jay Leno's Garage*, former host of *The Tonight Show*
Cliff Banks, Founder of The Banks Report
Jamie Butters, Editorial Page Editor of *Automotive News*
KC Crain, CEO of Crain Communications (*Automotive News*)
Paul Eisenstein, Founder of *Headlight News*
Bridget Fitzpatrick, Co-founder of *CBT News*
Jim Fitzpatrick, CEO and Co-founder *CBT News*
Matt Hardigree, Partner/Reporter with *The Autopian*
Tim Higgins, Columnist at *Wall Street Journal*
Phil LeBeau, Aviation and Automotive Correspondent for CNBC
Yossi J Levi, Founder, Car Dealership Guy
Ed Ludlow, Journalist and Broadcaster at *Bloomberg News*
John McElroy, Founder of Autoline TV
Keith Naughton, Auto industry reporter at *Bloomberg Business*
Nora Naughton, Automotive Reporter at *Business Insider*
Dan Neil, Automotive Columnist at *Wall Street Journal*
Edward Niedermeier, Director of Communications at Partners for Automated Vehicle Education (PAVE)
Alex Roy, Founder of Johnson & Roy Advisors, Autonocast, The Drive, Human Driving Association
David Shepardson, Automotive/Aviation Reporter for *Reuters*
Jason Torchinsky, Co-Founder of *The Autopian* and author of *Robot, Take the Wheel!*
David Tracy, Co-Founder of *The Autopian*
Craig Trudell, Global Automotive Editor for *Bloomberg*
Kevin Tynan, Senior Automotive Analyst at *Bloomberg Intelligence*
Bud Wells, Automotive reporter of *The Denver Post* and 6X book author

Two-wheel Mobility – Bikes, Scooters, Motorcycles
Allen Cowgill, Den Bicycle Lobby (DBL)
Jenn Dice, President and CEO of People for Bikes
David Edwards, President at Primal Wear
Bill Nesper, Executive Director of League of American Cyclists

Auto/Mobility Analysis, Testing and Rating Agencies
Sam Abuelsamid, Principal Research Analyst at Guidehouse Insights: Market
    Intelligence
Kyle Connor, Founder and CEO of Out of Spec Studio
Robby DeGraff, Manager Product and Consumer Insights of Auto Pacifica
Jake Fisher, Senior Reporter at Consumer Reports
Tyson Jominy, Vice President of JD Power
Ed Kim, President and Chief Analyst for Auto Pacifica
Roman Mica, Nathan Adlen and Andre Smirnoff, Leadership team at The Fast
    Lane Cars and Trucks
Avi Steinhauf, CEO of Edmunds

New Vehicle Retailers/Leaders – Some of the finest and most active in the country – representative sampling of high impact, positive messaging mobility leaders. Many more are out there continuously keeping wheels rolling.
John Adams, Heuberger Motors
Tom Abbott, Montrose Ford Nissan
Wes Abbott, Montrose Ford Nissan
Mickey Anderson, Baxter Automotive
Nancy Ariano, TruWest Chrysler Jeep Dodge Ram
Mike Aus, Durango Motors
Michael Aus, Durango Motors
Carrie Baumgart, Markley Motors
Anthony Banno, Phil Long Honda of Glenwood Springs
Matt Boone, Johnson Auto Plaza
Tim Biles, Pikes Peak Traveland
John Bowell, Shortline Buick GMC
Kent Bozarth, Ed Bozarth Chevrolet & Celebration Chevrolet
Mark Brady, Fisher Auto
Norman Braman, Braman Automotive
Bill Byerly, Peak Kia
Thom Buckley, Red Noland Automotive
Jeff Carlson, Glenwood Springs Ford and Past Chair of National Automobile
    Dealers Association
Zach Carlson, Glenwood Springs Ford
John Carroll, Ed Carroll Motor Company
Vince Cimino, Audi Colorado Springs
Bob Colbert, Emich Automotive
Scott Cook, Cook Chevrolet, Cook Subaru, Cook Ford

Tim Corwin, Corwin Automotive
Scott Crouch, Flatirons Subaru
Steve Dahle, Dahle Automotive and Red Rock Auto
Tom Daniels, Landmark Lincoln
Kevin Davis, Western Slope Automotive
Mike Dellenbach, Dellenbach Motors
Steve Dellenbach, Dellenbach Motors
Ivette Dominquez, Alpine Automotive Group of Colorado
Mike Drawe, Alpine Automotive Group of Colorado
Rob Edwards, Summit Automotive Partners
Scott Ehrlich, Toyota of Laramie
George Eidsness, Transwest
John Elway, Elway Dealerships
Fred Emich IV, Emich Automotive
Shawn Evans, Porsche of Littleton
Ben Faricy, The Faricy Boys
Paul Faricy, The Faricy Boys
Craig Fisher, Fisher Auto
Dan Fitzgerald, Pueblo Toyota
Fletcher Flower, Flower Motors
Fred Flower, Flower Motors
Fritz Flower, Flower Subaru
Jonathan Fowler, Fowler Holding Company
Jim Gebhardt, Gebhardt BMW
Joe Gebhardt, Davidson Gebhardt Chevrolet – Subaru of Greeley
Don Gerbaz, Berthod Motors
Bob Ghent, Ghent Chevrolet Cadillac
Alex Gillette, Summit Automotive Partners
George Gillette, Summit Automotive Partners
Courtney Goodgain, Arapahoe Kia
Amanda Gordon, GoJo Auto
Jeremy Hamm, Subaru of Pueblo & Solon Nissan
Darrin Hall, Christopher's Dodge World
Bill Hellman, Hellman Motors
Tim Hellman, Hellman Chevrolet
Gunnar Heuberger, Heuberger Motors
Blayne Johnson, Johnson Auto Plaza
Dick Johnson, Johnson Auto Plaza
Ryan Johnson, Johnson Auto Plaza
Dan Johnson, Len Lyall Chevrolet
William Johnson, Stevinson Lexus of Lakewood
Mike Jorgenson, Red Noland Automotive
William Knowles, Mike Maroone Automotive
Kevin Lamar, Western Slope Automotive
Greg Larson, Sill TerHar Motors

Chris Lenckosz, Empire Lakewood Nissan
David Liebowitz, Braman Automotive
Jon Lind, Burlington Ford
Jim Lyall, Len Lyall Chevrolet
Steve Maneotis, Victory Motors
Doug Markley, Markley Motors
Mike Maroone, Maroone Automotive
Mike Marquez, Fort Collins Kia
Todd Maul, Elway Dealerships
Scott McCandless, McCandless International
Doug McDonald, McDonald Automotive
Jim McDonald, McDonald Automotive
Michael McDonald, McDonald Automotive
Travis McKinley, Bonanza Ford
Robert McMann, Phil Long Ford Denver
Brandon Moreland, Fort Collins Chrysler Dodge Jeep Ram
Doug Moreland, Moreland Automotive
Jim Morehart, Morehart Murphy Regional Auto Center
Mark Miller, Ed Bozarth Chevrolet Buick
Aaron Mills, Schomp Automotive
Bonnie Murray, O'Meara Automotive
Mike Nixon, Grand Junction Chrysler Dodge Jeep Ram
Trent Olinger, Mercedes Benz of Loveland
Brian O'Meara, O'Meara Automotive
Evan O'Meara, O'Meara Automotive
Paige O'Meara, O'Meara Automotive
Tom Ondrako, Durango Motors
Mary Pacifico, Rickenbaugh Motors
Nick Pacifico, Rickenbaugh Motors
Lee Payne, Planet Honda, Planet Hyundai, Genesis of Golden
Jaymie Payne, Planet Hyundai
Michael Payne, Mountain Chevrolet
Mark Perez, Phil Long Ford Chrysler Jeep Dodge Ram Trinidad
David Perkins, Perkins Motors
Tom Perkins, Perkins Motors
Mark Peterson, Peterson Toyota/Volvo
Robert Penkhus, Bob Penkhus Motor Company
Mitch Pierce, Elway Dealerships
Steve Powers, McDonald Automotive
Vic Reem, Red Rock Auto
Ray Reilly, Red Noland Automotive
Lannie Ridder, Vince's GM Center
Greg Rowland, Morehart Murphy Regional Auto Center
Vince Schreivogel, Vince's GM Center
Tom Sellers, TruWest Chrysler Dodge Jeep Ram
Kevin Shaughnessy, Phil Long Dealerships

Scott Shimer, Fisher Auto
Mike Shaw, Mike Shaw Automotive
Michael Shaw, Mike Shaw Automotive
Brion Stapp, Stapp Interstate Toyota
Clinton Stapp, Stapp Interstate Toyota
Gregg Stone, Land Rover Flatirons
Jeff Silverberg, Elway Dealerships
Chris and Amy Smith, Infiniti of Denver
Carol Spradley, Spradley Motors
Wes Tabor, Honda of Greeley & Hyundai of Greeley
Jack TerHar, Sill TerHar Motors
Paul Tew, Crossroads Hyundai and Korf Automotive
Ross Turner, Turner Automotive
Ed Tynan, Tynan VW and Nissan
Sean Tynan, Tynan VW and Nissan
Matt Tynan, Tynan VW and Nissan
Tim Van Binsbergen, Mountain States Toyota
Carl Venstram, Arapahoe Kia, Grand Kia and Skyline Mitsubishi
Derek Vidmar, Vidmar Motors
Aaron Wallace, Schomp Automotive
Mike Ward, Mike Ward Automotive
Jay Weibel, Weibel Automotive
Bill Wilcoxson, Wilcoxson Buick GMC
Dan Wilson, Corwin Automotive
Phil and Ann Winslow, Winslow BMW
Tyrone Williams, Baxter Automotive
Brent Wood, Veteran automotive executive
Chase Yoder, Yoder Chevrolet
Warren Yoder, Weld County Garage
Beau Boeckmann, President of Galpin Motors
Marc Cannon, EVP of AutoNation
Chris Conroy, CEO of Holman Automotive
Jeff Dyke, COO of Sonic Automotive
Bryan Deboer, CEO of Lithia Motors
Jon Ferrando, CEO of Blue Compass RV
Jeff Dyke, COO of Sonic Automotive
Ernest Garcia III, CEO and Founder of Carvana
Melinda (Mindy) Holman, Chair of Holman Automotive
David Bruton Smith, CEO of Sonic Automotive
Daryl Kenningham, CEO of Group 1 Automotive
David Hult, CEO of Asbury Automotive
Robert Kurnick Jr., President of Penske Automotive
Roger Penske, CEO of Penske Automotive
Mike Manley, CEO of AutoNation

PART THREE
THE FUTURE

59

Looking at the amazing future of personal mobility

The future ain't what it used to be.
—*Yogi Berra*

The car is the ultimate mobile device. It's the ultimate in mobile computing. Think about it: It has a processor, it has a computer, it has a GPS, it has a camera, it has sensors. It's just like a smartphone on wheels.
—*Marc Andreessen*

For readers who have taken the time to learn how we got to this point in our quest for desirable, reliable, dependable, time-sensitive, personal mobility, congratulations. You have arrived at the future. For those who delved into the book only for a look at what is likely to come, here it is. Either way is fine. Without a doubt, regardless of how interesting the past was that got us here. When it comes to what's in store for us, fasten your seatbelts, the best is yet to arrive.

Here in the future section of the book, you should find a better understanding of hydrogen fuel cell technologies, both positive and negative, the amazing advancement of autonomous or driverless car technologies and the simply incredible developments in eVTOL (aka flying car or passenger drone) technology. Sooner than most people realize, all of these technology advancements will change—for the better—how humans traverse the earth.

The combination of electrification, hydrogen fuel cell technology, autonomous drive technology and flying cars, together will lead to safer travel, emission free travel, congestion-lessened travel and, thusly, more enjoyment and pleasure in travel. The extent to which the problematic issues referenced in Section Two of the book are truly still issues, even if improving, they will all significantly improve into the future.

Autonomous cars

The rapid advancement of technology has paved the way for autonomous vehicles, and numerous companies have entered the race to develop self-driving cars and trucks. These cutting-edge technologies have the potential to revolutionize transportation, offering increased safety, efficiency and convenience.

Here is a list of my ratings for the top 10 companies leading the way in autonomous car and self-driving truck technologies, from tenth to first, the best being #1.

10.) **Zoox:** A startup acquired by Amazon in 2020, Zoox is focused on developing a fully autonomous ride-hailing service. The company aims to design purpose-built autonomous vehicles that prioritize passenger safety and comfort. With Amazon's backing, Zoox has the needed resources available to accelerate its technology development and deployment.

9.) **Baidu:** Often referred to as the "Google of China", Baidu is a major player in autonomous driving. The company's Apollo platform provides a comprehensive suite of software and hardware solutions for autonomous vehicles. Baidu has successfully conducted autonomous driving tests on public roads and aims to deploy fully autonomous vehicles in the near future.

8.) **AutoPilot (Tesla):** Led by controversial entrepreneur Elon Musk, Tesla AutoPilot has made significant strides in autonomous driving. The company's Autopilot system is being used by Tesla owners, offering features such as advanced driver assistance and autonomous highway driving. Tesla's fleet, equipped with sensors and cameras, collects data to improve its self-driving capabilities continually. Unlike others setting an industry standard, Tesla's Autopilot does not use Lidar technology for operating autonomous driving. The lack of Lidar with Tesla concerns me and others, from a safety standpoint.

7.) **NVIDIA:** A renowned technology company, NVIDIA has made significant contributions to autonomous driving by providing powerful computing platforms. Its Drive platform offers high-performance hardware and software solutions for autonomous vehicles, enabling real-time processing of vast amounts of data. NVIDIA's partnerships with automakers and technology firms have solidified its position in the industry.

6.) **Mobileye (Intel):** Mobileye, an Intel subsidiary, is a global leader in computer vision-based advanced driver-assistance systems (ADAS). With a strong focus on safety, Mobileye's technology forms the foundation for many autonomous driving systems. The company's expertise in perception, mapping and machine learning has positioned it as a key player in the autonomous vehicle arena.

5.) **Blue Cruise (Ford):** BlueCruise allows for true hands-free driving on prequalified sections of divided highways called Hands-Free Blue Zones that make up over 130,000 miles of roads in North American as well as some roads in Europe and Asia. BlueCruise uses blue lighting on the digital instrument cluster to indicate when the vehicle is in a hands-free zone.

4.) **Aptiv:** Aptiv's success emanates from a strong, sustainable business and a portfolio of safe, green and connected solutions that are transforming the future of mobility and making the world a better place. Aptiv's path to becoming mobility's leading software and system's technologies integrator began years ago. Today Aptiv continues to position itself to be one of the best autonomous drive systems in the world.

3.) **Aurora:** Aurora is a driverless technology company that was founded in 2017 by Chris Urmson, Sterling Anderson, and Drew Bagnell. The company is based in Palo Alto, California, and has quickly become a leader in the development of self-driving technology for vehicles. Chris Urmson, the CEO of Aurora, is a former Google engineer who played a key role in the development of the company's self-driving car project, which eventually became Waymo.

2.) **Cruise (General Motors):** General Motors' Cruise is an autonomous vehicle subsidiary that has made remarkable progress in self-driving technologies. With a focus on ride-hailing, Cruise has been testing its autonomous vehicles on the streets of San Francisco and is working towards deploying a fully autonomous ride-hailing service. The company has attracted substantial investments from various stakeholders, solidifying its position in the industry. General Motors shut off testing of Cruise as a ride-share service following an (aforementioned) injury accident in San Francisco. Though Cruise is expected to be back up in limited operations soon.

1.) **Waymo (Google):** A subsidiary of Alphabet Inc. (Google's parent company), Waymo is widely regarded as the leader in autonomous driving. With extensive research and development, Waymo has achieved significant milestones, including the first fully autonomous ride-hailing service. Its advanced hardware and software systems have enabled it to accumulate millions of self-driven miles, making it a pioneer in the field.

Electric Vertical Takeoff and Landing Vehicles (eVTOLs): The Next Frontier in Personal Mobility

Since the dawn of aviation, mankind has dreamt of taking to the skies in personal flying machines. Over the years, this fascination with flying cars has been fueled by countless movies, TV shows and books depicting a world where commuting through the air is an everyday reality.

However, only in recent years has this vision started to materialize, with Vertical Takeoff and Landing Vehicles (VTOLs), also called passenger drones or flying cars, inching closer to becoming a viable mode of transportation. As we stand on the cusp of this revolutionary technology, it is crucial to examine the history, challenges and potential benefits of eVTOLs in shaping our future mobility.

In recent months I've attended several conferences and events that focused largely on this newly emerging industry. It's fascinating to see the level of energy, excitement and investment being generated by the early announcements from this prescient industry. One eVTOL stock was up by 230 percent on the year and has not yet made its first passenger flight. Is all the attention justified or will eVTOLs take their place in overhyped history somewhere between Segways and Google Glass?

History and pop-culture attraction in flying cars

The idea of flying cars has captured people's imagination for decades. Since the early 20th century, inventors, engineers and entrepreneurs have attempted to bring this dream to life. From movies such as *Chitty Chitty Bang Bang, Back to the Future Part II, Star Wars, Harry Potter and the Chamber of Secrets,* and television shows such as *The Jetsons, Futurama and Doctor Who* along with books such as *War of the Worlds, Blade Runner, The Hitchhiker's Guide to the Galaxy* and *The Hunger Games* (note, this is far from an exhaustive list), flying cars have played through American pop culture as our ultimate freedom of mobility and adventure.

Aircraft manufacturers including Airbus, Boeing, Embraer and Bell Helicopters and major airlines including Delta, United and Jet-Blue along with traditional automakers including Toyota, Stellantis, Hyundai, Ford, Volkswagen and Mercedes Benz are investing in developing eVTOLs, driven by the belief that they can revolutionize urban transportation and place their services at the heart of the fastest-developing segment of personal mobility.

Among the more than 200 reported eVTOL startups, these are my top 10 (alphabetical) trying to get off the ground in the race to the friendly blue skies of individualized, personal mobility.

- Archer Aviation: Archer Aviation is an impressive eVTOL start-up based in Silicon Valley, California is led by CEO and co-founder Adam Goldstein and is dedicated to urban air mobility. Archer's mission is to unlock the skies and reimagine how we live and spend time.

- Beta Technologies: Beta is a company based in Burlington, VT, which is working on sustainable aviation through development of emission free eVTOLs and CTOLs. Beta has worked with the federal government as well as investors in an effort to get eVTOLs off the ground. Its primary project for eVTOL service is Alia.

- Daroni Aerospace: Daroni Aerospace is a leading aerospace company specializing in the design, development and manufacturing of advanced aircraft and space systems including eVTOLs, with a focus on innovation and cutting-edge technology. Daroni is dedicated to pushing the boundaries of aerospace engineering and providing solutions for the future of air and space travel.

- Embraer Eve: The Embraer Eve is an all-electric vertical takeoff and landing (eVTOL) aircraft designed for urban air mobility. It features a sleek and futuristic design, zero emissions and the ability to transport passenger and cargo in urban environments with minimal noise and environmental impact.

- Hyundai Supernal: The Supernal eVTOL development is being driven by an extremely talented lineup of aviation, aeronautics, weather and urban design specialists. With Hyundai expertise in automotive manufacturing and the deep, professional lineup of other specialists, Supernal is blessed with the leadership team and expertise needed to be successful.

- Joby Aviation: Joby Aviation, headquartered in California, is developing an all-electric eVTOL aircraft. The company's prototype has undergone numerous successful test flights, showcasing its potential for urban air mobility. Joby aims to commence commercial operations by 2024, offering a sustainable and efficient mode of transportation.

- Lilium Jet: The Lilium Jet, developed by the German company Lilium, is an electric eVTOL aircraft with a sleek design. Based in Munich, Lilium aims to offer on-demand air travel services, operating between various cities. The Lilium Jet completed its maiden flight in 2019, and the company anticipates commencing commercial operations by 2025.

- Textron Nexus: Textron, owner of Bell, a renowned helicopter manufacturer, is developing the Nexus, an air taxi concept with eVTOL capabilities. The Nexus prototype, powered by a hybrid-electric propulsion system, offers a unique blend of vertical lift and forward flight. Textron anticipates commencing commercial operations by 2025, revolutionizing urban transportation.

- Volocopter: Volocopter, a German company, is focused on developing eVTOL aircraft for urban mobility. The Volocopter prototype has successfully completed manned test flights, demonstrating its potential for short-distance air travel. The company plans to launch its commercial operations in Singapore by 2025, offering a unique aerial transportation experience.

- Wisk Aero: Wisk Aero, originally a joint venture between Boeing and Kitty Hawk, aims to revolutionize urban mobility with its electric eVTOL aircraft. The Wisk Aero prototype, capable of vertical takeoff and landing, has been extensively tested in New Zealand. Wisk Aero envisions launching its air taxi service within the next few years, providing a sustainable and efficient alternative to traditional transportation.

Challenges loom ahead as eVTOLs startups work to become operational. These challenges and advantages are covered more thoroughly in a chapter ahead.

Despite the progress made in eVTOL technology to date, which is nothing short of incredible, significant challenges remain on the path to widespread adoption. Safety is a paramount concern, as integrating thousands of eVTOLs into the airspace without compromising existing air traffic is a complex task. And the need for efficient energy sources, noise reduction and reliable aerial autonomous systems pose significant engineering hurdles. Overcoming these challenges will require collaboration between governments, regulatory agencies and the private sector to establish robust standards and guidelines.

Advantages to society

The potential societal benefits of eVTOLs as a transportation and mobility option are immense. As problem solvers, eVTOLs could alleviate traffic congestion in densely populated urban areas, reduce travel times and improve productivity. They could also provide a lifeline in emergency situations, reaching remote or disaster-stricken areas quickly and efficiently. And eVTOLs have the potential to democratize transportation, providing access to remote regions and underserved communities. By utilizing electric propulsion, eVTOLs will contribute to reducing carbon emissions and combating climate change.

Most frequent questions:

- How high will eVTOLs fly? Answer: Expected between 500 feet to 1,500 feet, with physical capability to go much higher. Though FAA rulemaking will define more clearly.

- What will power eVTOLs? Answer: Most are expected to be Battery Electric Vehicles (BEVs). Though like the auto industry, hydrogen fuel cell and hybrid are being explored.

- Will eVTOLs fly autonomously? Answer: Yes, eventually. Though all the startup entrants anticipate using pilots until autonomous technology is perfected.

- How soon will eVTOLs first be operational? Answer: Several startups predict 2025. My estimate is closer to 2030 for significant growth in their travels and services. At first they will use very limited pre-determined routes and grow from there.

- If eVTOLs fly autonomously, how do they avoid traffic congestion in the air? Answer: Based on autonomous drone shows with 2,000 to 5,000 participating with finely choreographed movement routines, similar autonomous safety technology will be able to ensure safe airspace even in a crowded traffic scene.

The prospect of eVTOLs becoming a reality is no longer a distant dream and a tangible realistic possibility. The history and interest in flying cars, coupled with their portrayal in popular culture, has cemented their place in our collective consciousness. While challenges remain, the advantages to society are undeniable. eVTOLs have the potential to reshape urban transportation, reduce traffic fatalities, expedite travel and usher in a new era of expedient connectivity. As we navigate the roadblocks ahead, it is crucial to foster collaboration and innovation, ensuring the promise of eVTOLs becomes a reality for all. The sky is no longer the limit; it is the horizon of a new era in energized mobility freedom.

60

Aurora

Pioneering driverless technology for future cars and trucks

"Aurora is pushing the boundaries of what's possible with driverless technology and is poised to revolutionize the transportation industry."
—Tim Cook, CEO of Apple

Aurora is an autonomous vehicle technology company that is revolutionizing the automotive industry by developing cutting-edge driverless car and driverless truck technology. Founded in 2017, Aurora has quickly emerged as one of the industry leaders, attracting significant investments and partnerships. This chapter aims to delve into the company's founding, its location, notable investors and the progress it has made in bringing its groundbreaking technology to market.

Aurora was founded in 2017 by three prominent figures in the autonomous vehicle industry: Chris Urmson, Sterling Anderson and

Drew Bagnell. Urmson, the CEO, was formerly the CTO of Google's self-driving car project, which later became Waymo. Anderson, Aurora's chief product officer, previously worked as the director of Tesla's Autopilot program. Bagnell, the company's Chief Technology Officer, was a founding member of the Uber Advanced Technologies Group.

The founders shared a vision of creating a company that could bring safe, efficient and accessible autonomous vehicle technology to market. Leveraging their vast experience and expertise, Aurora set out to develop a full-stack autonomous driving system that could be deployed across a range of vehicles, including cars and trucks. Their earliest progress on testing in real time settings was for trucks. Specifically, over the road semi-trucks.

Aurora is headquartered in Palo Alto, California, a hub for technological innovation in the heart of Silicon Valley. The company also has operations in Pittsburgh, Pennsylvania, a city renowned for its expertise in robotics and autonomous systems, due largely to the early studies and investment by Carnegie-Mellon University in Pittsburgh. By having a presence in both locations, Aurora can tap into the talent pools and research facilities of these two prominent tech hubs, further strengthening its position in the autonomous vehicle industry. Aurora's facilities include state-of-the-art engineering labs, simulation centers and testing grounds. These resources enable the company to develop and refine its technology, ensuring its readiness for real-world applications.

Notable investors

Aurora has attracted significant investments from both traditional automotive giants and technology-focused venture capital firms. These investments showcase the industry's confidence in Aurora's technological prowess and its potential to shape the future of transportation. Some notable investors include:

- Amazon
  In early 2019, Amazon led a $530 million funding round for
  Aurora. This investment highlighted Amazon's interest in
  leveraging autonomous vehicle technology for its logistics
  and delivery operations.

- Sequoia Capital
  One of the leading venture capital firms in Silicon Valley,
  Sequoia Capital has been a key investor in Aurora. Their i
  investments have provided the necessary financial backing
  for Aurora's ambitious research and development efforts.

- Hyundai Motor Group
  In 2020, Hyundai Motor Group invested $600 million in
  Aurora, solidifying their partnership. This collaboration
  aims to accelerate the development of Hyundai's
  autonomous vehicles by integrating Aurora's technology.

- Uber
  Aurora and Uber forged a partnership in 2020, with Uber
  investing $400 million in the company. This collaboration
  aims to combine Uber's ride-share platform with Aurora's
  autonomous technology, paving the way for a driverless
  future for ride-sharing services. These investments and
  partnerships underline the confidence industry leaders
  have in Aurora's technology, positioning the company as
  a frontrunner in the race towards autonomous vehicles.
  It should be noted, though, that Uber recently contracted
  with Waymo One for ride-share services in Phoenix and
  other Waymo-approved cities. So, I'm not sure where that
  will put Uber's earlier investment with Aurora. I imagine
  neither Waymo nor Aurora will have exclusive technology
  access to Uber's platform.

Aurora has made significant strides in advancing its driverless technology, focusing on safety, scalability and commercial viability. The company's approach involves developing a full-stack solution that encompasses hardware, software as well as data services. This has enabled seamless integration into a wide range of vehicle platforms.

Aurora's technology is already being tested on public roads, accumulating millions of autonomous miles in various weather and traffic conditions. The company's fleet of test vehicles, equipped with advanced sensors and perception systems, allows for continuous data gathering and refinement of its algorithms.

Aurora's technology demonstrates scalability and adaptability, making it well-suited for a wide range of commercial applications. The company's approach of building partnerships with automakers and transportation companies allows it to integrate its autonomous technology seamlessly into existing vehicle platforms. This collaborative approach enables Aurora to leverage the expertise and manufacturing capabilities of established industry players, accelerating the commercial deployment of their technology. Aurora's adaptable software and hardware architecture can be customized to suit specific industry requirements, expanding its potential for commercial viability across various sectors.

The commercial viability of Aurora's autonomous car and truck technology is closely tied to the regulatory framework and public acceptance. As governments worldwide continue to develop regulations for autonomous vehicles, Aurora has been actively engaged in collaborating with policymakers to ensure the safe and responsible deployment of its technology. By addressing regulatory concerns and building public trust through extensive testing and transparency, Aurora aims to pave the way for widespread adoption of autonomous vehicles, thus enhancing its commercial viability.

From my vantage point, as an observer of the autonomous car technology industry, Aurora is well-positioned to be a provider of this all-important technology. I think it is ahead of everyone, with the possible exception of Waymo, in autonomy for trucks. For cars I would rank Aurora in "Tim's Top Ten" at Number Three—still a very impressive start with proven, engaged and time-tested leadership for the company. For over the road semi-trucks, I rank Aurora as Number One in the early stages of autonomous technology startups.

Aurora's autonomous car and truck technology have the potential to revolutionize transportation across multiple sectors. From ride-hailing and ride-sharing services to long-haul trucking and public transportation, the commercial applications are vast. As autonomous driving technology continues to evolve and mature, it is likely to become an integral part of our daily lives, transforming the way we commute and transport goods. With strategic partnerships, a relentless focus on safety and a commitment to scalability, Aurora's technology is poised for long-term commercial success.

Aurora's autonomous car and truck technology showcases immense commercial viability due to its emphasis on safety, reliability, efficiency and adaptability. By addressing regulatory concerns and building public trust, Aurora seems well-positioned to capitalize on the growing demand for autonomous vehicles. As the technology continues to advance and gain wider acceptance, it is expected to revolutionize transportation across various sectors, making Aurora a key player in the autonomous vehicle industry. And first, so far, among over-the-road's semi-trucks.

61

Cruise

General Motors' Cruise remains on path toward fully-autonomous cars

"GM's Cruise driverless cars are a testament to the company's commitment to advancing the future of transportation."
—Sundar Pichai, CEO of Alphabet

General Motors' Cruise is a ride-share service founded in 2013 by Kyle Vogt and Dan Kan. The company was acquired by General Motors (GM) in 2016 for a reported price of $1 billion. Since then, Cruise has become one of the most talked-about autonomous vehicle companies in the world. Cruise is also one of the most real-life tested autonomous ride-share services in the world.

Cruise is headquartered in San Francisco, and has operations in several other cities, including Phoenix, Seattle, Raleigh and Austin. The company is focused on developing self-driving cars that can operate safely and efficiently on public roads. Cruise's ultimate goal is to create a fully autonomous ride-share service that can compete with traditional taxi and ride-share services.

The founders of Cruise, Vogt and Kan, were initially entrepreneurs with a track record of building successful companies. Vogt was the founder of Twitch, a popular video game streaming platform that was acquired by Amazon for $970 million in 2014. Kan was the founder of Exec, an on-demand cleaning and errand service that was acquired by Handy in 2014.

GM initially retained the services of Vogt and Kim in leadership roles for Cruise. Together, they built a team of engineers and designers working to develop the technology that will power Cruise's autonomous vehicles. The team included experts in machine learning, computer vision, robotics and more.

After a Cruise vehicle was involved in an accident in San Francisco, that arguably should have been avoided or have resulted in a less severe outcome, General Motors then took all Cruise autonomous robo-taxis out of service and later terminated most all of the top tier of Cruise leadership. The next steps' plan is to re-enter service slowly, carefully and in limited/reduced markets. GM now plans to invest less new capital in Cruise as it anticipates a slower new rollout of its ride-share services.

One of the key advantages Cruise had over other autonomous vehicle companies is its partnership with General Motors. GM is one of the largest automakers in the world, with decades of experience in designing and manufacturing cars. This relationship provides Cruise access to the resources and expertise of a major automaker, a significant advantage in the highly competitive world of autonomous vehicles.

Cruise has also received significant investment from other companies, including SoftBank and Honda. SoftBank invested $2.25 billion in Cruise, while Honda invested $2.75 billion, both in 2018. These investments have helped fund Cruise's research and development efforts and have allowed the company to expand its operations.

Despite its significant resources and talented team, Cruise has faced several challenges in developing its autonomous vehicle technology. One of the biggest has been ensuring that the vehicles can operate safely and efficiently in a wide range of conditions. This includes dealing with unpredictable weather, navigating through dense urban environs and responding to unexpected obstacles on the road.

To address these challenges, Cruise has been testing its autonomous vehicles extensively in real-world conditions. The company has a fleet of over 600 self-driving cars being tested on public roads in several cities. These tests have allowed Cruise to gather data and refine its technology, which has helped to improve the safety and efficiency of its autonomous vehicles.

Cruise has been working closely with local, state and federal regulators and policymakers to ensure that its autonomous vehicles comply with all relevant laws and regulations. This includes working with the NHTSA and the US Department of Transportation (US-DOT) to develop safety standards for autonomous vehicles.

Despite these challenges, Cruise has truly made significant progress in developing its autonomous vehicle technology. The company has recently launched a fully autonomous ride-share service in San Francisco. This service, initially available to Cruise employees and their families, was soon expanded to the general public.

When fully operational, Cruise's ride-share service will compete with traditional taxi and ride-share services, such as Uber and Lyft, though without drivers, being fully navigated through artificial intelligence. However, Cruise's autonomous vehicles will have several advantages over the traditional driver-provided ride-share services. For example, the vehicles will be able to operate 24/7 without the need of a human driver, which will allow the company to provide a more efficient and cost-effective service.

It's anticipated that Cruise's autonomous vehicles will also be able to provide a safer service than traditional taxi and ride-share services. According to the NHTSA, 94% of car accidents are caused by human error. By removing the human driver from the equation, Cruise's autonomous vehicles will be able to significantly reduce the risk of accidents and improve overall safety.

In addition to its ride-share service, Cruise is exploring other applications for its autonomous vehicle technology. The company is developing a delivery service that will use autonomous vehicles to transport goods and packages. This service could provide a more efficient and cost-effective alternative to traditional delivery services, such as UPS and FedEx.

The success of Cruise has been impressive, but the company faces significant challenges ahead. One of the biggest challenges will be scaling its autonomous vehicle technology to meet the demands of a large ride-share service. This will require significant investment in research and development, as well as the development of new manufacturing processes and supply chains.

Cruise will also need to continue working closely with regulators and policymakers to ensure that its autonomous vehicles comply with all relevant laws and regulations. This will be particularly important as the company expands its ride-share service to more cities and regions.

Despite these new challenges Cruise faces, in my humble opinion (IMHO), the longer term future remains bright for Cruise. The company has a talented mid-level engineering team, enough resources and the all-important strong partnership with General Motors. With additional continued investment and development, Cruise has the potential to help revolutionize the transportation industry and become a major player in the world of autonomous vehicles and robo-taxi, ride-share services.

Additionally, the technology behind the Cruise robo-taxi cars in ride-share services is similar to what General Motors is using for customer-driven cars. That software and vehicle option is called Supercruise and is available today in several GM models sold under its traditional vehicle brands, including Cadillac, Buick, GMC and Chevrolet. By most reviews, Supercruise is considered one of the best fully autonomous vehicle technologies in the new car market today. Over time Supercruise autonomous driving technology will continue to get better, more reliable, more trusted and even more expected.

Based on my eight-year industry observation of autonomous, self-driving, driverless (using all three common synonyms) technologies along with investment, infrastructure and track record, I rate Cruise among the few in development currently that is most likely to succeed. In fact, I still rate Cruise as Number Two (#2) most likely to succeed, immediately behind Waymo, the autonomous car company founded by Google and led in its early years by the legendary John Krafcik.

62

Waymo

Absolute early leader in autonomous drive technology

"Our ultimate goal is to bring full self-driving technology to the world, making it safe and easy for people and things to get where they're going."
—Founding Waymo CEO John Krafcik

Waymo, an autonomous vehicle company operating since 2009, was originally part of Google's self-driving car project, which was created to develop autonomous vehicles that could operate on public roads. In 2016, the company became a standalone subsidiary of Alphabet Inc., Google's parent company.

Founders

The founders and early leaders of Waymo are John Krafcik, Anthony Levandowski and Chris Urmson. Krafcik joined Waymo as CEO in 2015, bringing with him extensive experience in the automotive industry. Prior to joining Waymo, Krafcik worked at Hyundai Motor America, where he served as president and CEO. Levandowski, on the other hand, was one of the early engineers on the Google self-driving car project and is credited with developing many of the key technologies that enable autonomous driving. Urmson, also an early member of the self-driving car project, left Waymo in 2016 to start his own autonomous vehicle company, Aurora.

Waymo's history can be traced back to the DARPA Grand Challenge, a competition sponsored by the U.S. government to develop autonomous vehicles. In 2005, Google co-founder Larry Page saw a demonstration of the technology developed for the Grand Challenge and was impressed by the potential of autonomous vehicles. Google began investing in autonomous vehicle technology, and in 2009 the company launched the self-driving car project.

Since then, Waymo has been developing and testing its autonomous vehicle technology on public roads. The company has logged over 20 million miles of autonomous driving on public roads and has conducted extensive testing in simulation. Waymo's vehicles use a combination of sensors, such as lidar, radar and cameras, to perceive their surroundings and navigate the roads. The vehicles are also equipped with advanced software and machine-learning algorithms that enable them to make decisions in real-time.

First ride-share opportunity

In 2018, Waymo launched a commercial ride-hailing service in Phoenix, called Waymo One. The service allows users to hail a ride in one of Waymo's autonomous vehicles using a smartphone app. Waymo One is available to a limited number of users in the Phoenix area, and the company has begun to expand to other cities, including San Francisco, Austin and Los Angeles.

For the past five years or so, I have spent quite a few weekends and holidays in the Phoenix area between October and April. Over those years, I have witnessed a significant increase in sightings of Waymo One cars out doing autonomous car testing in Phoenix and neighboring communities including Tempe, Gilbert, Chandler, Mesa and Scottsdale.

In November 2023, I saw the most Waymo test cars I have ever seen while pedaling my bicycle around the Phoenix metro area on a 50-mile cycling ride. Note, too, that I am also seeing more Cruise autonomous cars on the roads being tested as well. Cruise just recently announced it is shutting down passenger operations until further notice. Waymo has not had to do that. Waymo, for now, is a step or more ahead of Cruise.

Waymo's plans for the next decade include further expanding its ride-hailing service, as well as developing autonomous vehicles for other industries, such as logistics and delivery. The company is also working on developing technology for autonomous trucks, which it believes has the potential to transform the trucking industry. Waymo has already begun testing autonomous trucks in Arizona and plans to expand its testing to other states in the coming years.

One of the biggest challenges facing Waymo and other autonomous vehicle companies is the regulatory environment. Regulations governing autonomous vehicles vary by state, and there is no federal framework for regulating these vehicles. Waymo has been working with policymakers and regulators to develop a framework that will enable autonomous vehicles to operate safely on public roads. The company has also been working with other stakeholders, including insurance companies and law enforcement, to ensure that autonomous vehicles are integrated safely into the existing transportation system.

Challenges toward adoption

Another challenge facing Waymo is the cost of developing and deploying autonomous vehicles. The technology required for autonomous driving is still relatively expensive, and it may be some time before it becomes cost-effective for widespread deployment. Waymo has been working to drive down the cost of its technology through partnerships with suppliers and other companies in the supply chain. The company has also been exploring new business models, such as leasing its vehicles to ride-hailing companies, to help offset the cost of developing and deploying its technology.

Despite these challenges, Waymo has made significant progress in developing its technology and bringing it into this very new market. The company's ride-hailing service, Waymo One, represents a major milestone in the development of autonomous vehicles, and has the potential to transform the transportation and mobility industry. As Waymo continues to develop its technology and expand its operations, it will be interesting to see how the company navigates the challenges ahead and helps to shape the future of transportation.

Winning at Waymo

Certainly, Waymo is one of the leading autonomous vehicle companies that has been developing and testing its technology for over a decade. John Krafcik, an early CEO of Waymo, is a longtime friend of mine. He is a rising star and has been all of his adult life. I credit John with being the brains and thought leader behind the rise of Hyundai North America. Between 2008 and 2014, Hyundai more than doubled its sales in the US, from about 600,000 to more than 1.2 million. It was one of only three car brands in the US to grow sales during the 2008-2009 recession. That was largely the responsibility of John Krafcik. John was also an excellent early leader of Waymo and should be credited with the important lift and assist to get where Waymo is today.

As for Waymo's viability as an autonomous car technology manufacturer, I personally rate it as Number One (#1) and most likely to succeed, based on American companies in this arena so far. I think Waymo and Cruise are three to five years ahead of everyone else. Watch Waymo help set the standard for what I believe autonomous car technology should look like, how it should operate and what other companies should do to help ensure safer roads through best use cases of Autonomous Intelligence (considered the most advanced type of artificial intelligence—AI).

63

Hydrogen

Hydrogen fuel cell technology – future power source for cars and trucks

The automotive industry has been exploring various alternative fuel options to reduce greenhouse gas emissions and dependency on fossil fuels. Hydrogen fuel cell technology has emerged as a promising solution, offering clean energy and zero-emission vehicles. This chapter aims to provide an in-depth analysis of hydrogen fuel cell technology for both cars and heavy trucks, focusing on leading manufacturers including Toyota, Ford, Porsche, Hyundai, BMW, Honda, General Motors, Mercedes Benz and Nikola. Additionally, it will examine the prospects for the viability of hydrogen fuel cell vehicles in both primary sectors, cars and heavy trucks.

Hydrogen fuel cell technology for cars

Hydrogen fuel cell vehicles (FCVs) use hydrogen gas to produce electricity through a chemical reaction with oxygen, generating power to propel the vehicle. The only byproduct of this reaction is water vapor, making FCVs an eco-friendly alternative to traditional internal combustion engine vehicles. Several automobile manufacturers have heavily invested in hydrogen fuel cell technology for cars. Toyota is at the forefront of FCV development with its flagship model, the Toyota Mirai, one of the most successful hydrogen-powered vehicles in this very young market. Toyota's commitment to hydrogen fuel cell technology is evident in its investments in hydrogen infrastructure and partnerships with other stakeholders.

Hyundai is also a major player in the hydrogen car market, with its Hyundai Nexo, which offers extended range and advanced safety features. BMW has been actively researching and developing hydrogen fuel cell technology, aiming to integrate FCVs into its future vehicle lineup. Honda has also made significant strides in this field, with the Honda Clarity Fuel Cell being one of the world's first commercially available FCVs. GM, through its partnership with Honda, plans to develop next-generation fuel cell systems, demonstrating GM's commitment to FCV technology. Ford has been working on hydrogen fuel cell technology for more than two decades and is advancing tests on hydrogen-powered vehicles.

Prospects for viability of hydrogen fuel cell cars

While hydrogen fuel cell technology for cars has shown great potential, there are several challenges to overcome for widespread adoption. The main obstacle is the lack of an extensive hydrogen refueling infrastructure, limiting the practicality and convenience of FCVs. However, governments and private entities are investing in hydrogen infrastructure development, with early initiatives such as the California Hydrogen Highway Network and various European projects.

Another challenge is the cost of producing and storing hydrogen. Currently, hydrogen production relies heavily on fossil fuels, which undermines its environmental benefits. However, advancements in renewable energy sources, such as solar and wind power, offer a potential solution to produce green hydrogen, making FCVs even more sustainable.

Despite these challenges, hydrogen fuel cell cars offer several advantages. They have longer driving ranges and faster refueling times compared to BEVs. FCVs also eliminate range anxiety, as they do not rely on a limited battery capacity. Further, the scalability of hydrogen fuel cell technology allows its application to benefit various vehicle types, including SUVs and sedans.

Here is a summary of activity and efforts through leading automakers, by brand, who are stepping up on hydrogen fuel cell technology and what they are focused on so far:

- Toyota
  - The company believes that hydrogen fuel cell electric powertrains have the potential to be a sustainable and efficient solution for various applications.
  - To explore the viability of using hydrogen fuel cells in commercial vehicles, Toyota is conducting extensive research and development activities. The company is investing in the development of advanced fuel cell stacks, hydrogen storage systems, and overall system optimization.
  - Toyota is collaborating with various partners, including commercial fleet operators and government agencies, to conduct real-world testing and gather data on the performance and feasibility of hydrogen fuel cell vehicles in different commercial applications.

- Toyota has already introduced hydrogen fuel cell technology in its Mirai sedan and is now working on expanding its lineup of commercial vehicles powered by fuel cells. The company recently announced plans to develop a hydrogen fuel cell-powered commercial truck called the "Project Portal" for heavy-duty applications.
- Through these efforts, Toyota is actively exploring the potential of hydrogen fuel cell technology for commercial vehicles and working towards a sustainable and zero-emission future in transportation.

- Ford
  - Partnership with Ballard Power Systems: Ford has collaborated with Ballard Power Systems, a leading provider of fuel cell solutions, to develop fuel cell technology for its commercial vehicles. The partnership aims to enhance fuel cell durability and reduce costs, making it a more viable option for commercial applications.
  - Proving ground testing: Ford has conducted extensive testing of its hydrogen fuel cell prototypes at its Michigan Proving Ground. These tests help evaluate the performance, durability, and efficiency of the fuel cell systems, ensuring they meet the requirements of commercial vehicle operations.
  - Research on commercial vehicle applications: Ford is conducting research to understand the specific needs and requirements of commercial vehicle operators in various sectors, such as delivery, public transportation, and logistics. This research helps in tailoring hydrogen fuel cell technology to address the unique challenges faced by commercial vehicles.

- Investment in hydrogen infrastructure: Ford recognizes
  that the success of hydrogen fuel cell vehicles depends
  on the availability of a robust infrastructure. Therefore,
  the company has invested in partnerships and
  collaborations to support the development of hydrogen
  refueling stations. This investment aims to create a
  more accessible and reliable hydrogen infrastructure
  network for commercial vehicle operators.
- Collaboration with other automakers: Ford is a member
  of the Hydrogen Council, a global initiative that brings
  together leading automakers, energy companies, and
  other stakeholders to promote hydrogen as a clean
  energy carrier. Through this collaboration, Ford
  contributes to the collective efforts of the industry to
  advance hydrogen fuel cell technology. By engaging
  in these activities, Ford is actively researching the
  viability of hydrogen fuel cell engines for its commercial
  vehicles, with the goal of providing sustainable and
  efficient transportation solutions.

- BMW
  - The company has partnered with Toyota to work on
    joint research projects since 2013. The aim of this
    collaboration is to enhance the development and
    production of fuel cell systems for various applications,
    including commercial vehicles.
  - BMW has also announced plans to launch a small series
    of hydrogen fuel cell-powered vehicles by 2022, with a
    focus on commercial applications.
  - The company aims to test the technology in real-world
    conditions to assess its viability and potential integration
    into their commercial vehicle lineup. Furthermore, BMW

is investing in its own research and development facilities to advance hydrogen fuel cell technology.
- BMW has a dedicated hydrogen fuel cell research center in Munich, Germany, where engineers are working on developing and optimizing fuel cell systems for commercial vehicles. Overall, BMW's commitment to research, partnerships, and dedicated research facilities showcases its efforts to explore the viability of hydrogen fuel cell engines for commercial vehicles.

- Hyundai
  - Hydrogen fuel cell vehicle lineup: Hyundai already offers hydrogen fuel cell vehicles, including the Hyundai NEXO SUV and the Hyundai Xcient Fuel Cell truck. These vehicles serve as a crucial testing ground for the technology and help gather data for further research.
  - Investment in R&D: Hyundai has made significant investments in research and development to enhance the performance, efficiency, and durability of hydrogen fuel cell systems. This includes improving the fuel cell stack, reducing costs and increasing overall vehicle range.
  - Collaboration with partners: Hyundai collaborates with various partners to accelerate hydrogen fuel cell technology development. For instance, the company works closely with Cummins, a global power leader, to jointly develop and commercialize fuel cell powertrains for commercial vehicles.
  - Hydrogen infrastructure: Hyundai recognizes the importance of building a comprehensive hydrogen infrastructure to support the widespread adoption of fuel cell vehicles. The company has been actively involved in developing hydrogen refueling stations and promoting

collaboration among stakeholders to establish a robust hydrogen ecosystem.

- Future plans: Hyundai has ambitious plans for hydrogen fuel cell commercial vehicles. By 2028, the company aims to introduce a range of fuel cell systems for various commercial applications, including trucks, buses, and trains. These efforts are part of Hyundai's commitment to achieving a carbon-neutral society and promoting sustainable transportation solutions.
- Hyundai is dedicated to researching and advancing the viability of hydrogen fuel cell engines for its commercial vehicles, with a focus on improving technology, expanding infrastructure, and collaborating with key stakeholders.

- Mercedes
    - Joint Ventures: Mercedes-Benz has partnered with Volvo Trucks to form a joint venture called "Cellcentric." This collaboration aims to accelerate the development, production, and commercialization of fuel cell systems for heavy-duty commercial vehicles.
    - Prototypes and Trials: Mercedes-Benz has developed several hydrogen fuel cell prototypes for commercial vehicles, including the Mercedes-Benz GenH2 Truck. These prototypes are used for testing and gathering data on the performance and feasibility of hydrogen fuel cell technology in real-world conditions.
    - Investment in Infrastructure: Mercedes-Benz is investing in the development of a hydrogen infrastructure to support the deployment of fuel cell vehicles. This includes collaborations with energy companies and the establishment of hydrogen refueling stations.
    - Research and Development: Mercedes-Benz is continuously investing in research and development to

enhance the efficiency, range, and durability of hydrogen fuel cell technology. They are exploring advancements in fuel cell stacks, hydrogen storage, and overall system integration.

– Collaboration with Industry and Government: Mercedes-Benz collaborates with other industry players, research institutions and government bodies to share knowledge, promote standardization, and drive the growth of hydrogen fuel cell technology. These efforts reflect Mercedes-Benz's commitment to exploring sustainable alternatives for commercial vehicles, with a specific focus on hydrogen fuel cell technology.

• General Motors and Honda
  – Partnership with GM and Honda: Together they have collaborated to jointly develop next-generation fuel cell systems and hydrogen storage technologies. This partnership aims to reduce costs and accelerate the commercialization of fuel cell vehicles. The companies have invested $2.5 billion in their joint venture, Fuel Cell System Manufacturing (FCSM), to enhance fuel cell technology.
  – Hydrogen Fuel Cell Truck: GM/Honda is working on developing a commercial hydrogen fuel cell truck. In collaboration with the U.S. Army Tank Automotive Research, Development and Engineering Center (TARDEC), GM/Honda is developing a prototype fuel cell truck platform called the Chevrolet Colorado ZH2. This project helps GM/Honda gather valuable data and insights to further refine and optimize fuel cell technology for commercial use.
  – Fuel cell research facility: GM/Honda operates a Fuel Cell Research and Development Center in Pontiac, Michigan.

This facility focuses on developing advanced fuel cell technologies and improving their efficiency, durability and performance. The research center plays a crucial role in GM/Honda's efforts to explore the viability of hydrogen fuel cell engines for commercial vehicles.

- Fleet testing: GM/Honda has conducted various fleet tests to evaluate the performance and real-world viability of hydrogen fuel cell vehicles. For example, the Chevrolet Equinox Fuel Cell, a hydrogen-powered SUV, has been utilized in fleet trials to gather data and feedback from real-world driving conditions. These tests help GM/ Honda understand the requirements and challenges of integrating fuel cell technology into commercial vehicles.

- Collaboration with the U.S. Department of Energy: GM has partnered with the U.S. Department of Energy (DOE) for various research projects related to fuel cell technology. Through collaborations with the DOE's National Renewable Energy Laboratory (NREL) and other institutions, GM aims to advance the understanding and application of hydrogen fuel cells in commercial vehicles.

- Overall, GM and Honda are actively engaged in researching and developing hydrogen fuel cell technology for commercial vehicles. Through partnerships, research facilities, fleet testing, and collaborations with government organizations, GM/Honda is working towards the viability and commercialization of hydrogen fuel cell engines.

- Porsche
  - Within this framework, Porsche is specifically involved in the development of fuel cell systems.
  - The company aims to leverage its expertise in electric

drivetrains and combine it with hydrogen fuel cell technology to create sustainable and efficient powertrains for its commercial vehicles.

– Porsche has also joined several collaborative projects to advance fuel cell technology. For instance, the company is a partner in the "H2 Mobility" initiative in Germany, which aims to establish a hydrogen infrastructure across the country.

– Porsche is also part of the "Hydrogen Council," the global initiative that promotes hydrogen as a key solution for decarbonizing transportation and other sectors. By actively participating in these initiatives and investing in research and development, Porsche is demonstrating its commitment to exploring the feasibility of hydrogen fuel cell engines for its commercial vehicles.

Hydrogen fuel cell technology for trucks

The potential for hydrogen fuel cell technology extends beyond cars, with heavy-duty trucks a key area of focus. Companies like Nikola have recognized the advantages of hydrogen fuel cell technology for long-haul trucks. Nikola's hydrogen fuel cell trucks offer extended range, reduced emissions, and faster refueling times compared to conventional diesel-powered trucks. Additionally, they provide a quieter and more comfortable driving experience.

Other manufacturers, including Toyota and Hyundai, are exploring the application of hydrogen fuel cell technology in the trucking industry. Toyota's Project Portal and Hyundai's Hydrogen Mobility Solution are examples of initiatives aimed at developing fuel cell trucks. These advancements in hydrogen fuel cell technology for trucks when implemented will contribute to the decarbonization of the transportation sector, reducing emissions from heavy-duty vehicles.

Prospects for viability of hydrogen fuel cell trucks

The prospects for hydrogen fuel cell trucks (FCTs) are promising, given their potential to address the challenges faced by electric trucks, such as limited range and longer charging times. Scalability and quick refueling capabilities will beessential ahead of widespread adoption of FCTs.

Fuel cell technology: advantages and disadvantages as a power source for cars and heavy trucks

As the global concern for reducing greenhouse gas emissions and combating climate change intensifies, the search for clean and sustainable energy sources has become a top priority. Hydrogen fuel cell technology has emerged as a promising alternative to conventional fossil fuels, offering numerous advantages and disadvantages as a power source for cars and heavy trucks.

Here we explore the benefits and drawbacks of hydrogen fuel cell technology and highlight investments made by existing car, oil and truck manufacturers, as well as other stakeholders, in this cutting-edge technology.

Advantages of hydrogen fuel cell technology

- Zero emissions: One of the most significant advantages of hydrogen fuel cells is their ability to produce zero greenhouse gas emissions. As mentioned earlier, hydrogen gas, when combined with oxygen in a fuel cell, generates electricity, releasing only water vapor as a byproduct. This makes hydrogen fuel cells a clean and environmentally friendly alternative to internal combustion engines.

- Energy efficiency: Hydrogen fuel cells boast higher energy efficiency compared to traditional combustion engines.

Fuel cells convert chemical energy directly into electrical energy, without the need for intermediate mechanical processes. Consequently, fuel cell-powered vehicles can achieve higher mileage per kilogram of hydrogen, resulting in improved energy utilization.

- Quicker refueling: Unlike electric vehicles that require hours to recharge their batteries, hydrogen fuel cell vehicles can be refueled in a matter of minutes, similar to conventional gasoline or diesel vehicles. This advantage reduces or eliminates range anxiety and offers a more convenient solution for long-distance travel or heavy-duty applications.

- Versatility: Hydrogen fuel cell technology is highly versatile and can be implemented in various vehicle types, including cars, buses and heavy trucks. This adaptability allows for a broader range of applications and a seamless integration into existing transportation infrastructure.

Disadvantages of and challenges to hydrogen fuel cell technology

- Infrastructure challenges: A significant disadvantage of hydrogen fuel cell technology is the lack of an extensive refueling infrastructure. Building a comprehensive network of hydrogen refueling stations requires substantial investments and time. Without a well-developed infrastructure, the adoption of hydrogen-powered vehicles may face limitations, hindering their widespread usage.

- Cost and production: The production, storage and transportation of hydrogen present challenges in terms of cost and scalability. Currently, hydrogen fuel cells are more expensive to produce than internal combustion engines or battery systems. Additionally, the production of hydrogen primarily relies on natural gas reforming, which contributes to carbon emissions during its creation. Developing cost-effective and sustainable methods for hydrogen production is crucial to overcome this obstacle.

- Limited hydrogen availability: Although hydrogen is the most abundant element in the universe, it is primarily found in compound form on Earth. Extracting hydrogen from these compounds, such as water or hydrocarbons, requires energy-intensive processes, reducing the overall efficiency of the fuel cell system. The availability of hydrogen may be limited in certain regions, making it less viable as a global energy solution.

A new, yet little-known wrinkle: rising water vapor in the atmosphere: a catalyst for climate change and implications for hydrogen fuel cell vehicles

The issue of climate change has become one of the most pressing challenges of our time. While the causes of global warming are multifaceted, the role of rising water vapor in the atmosphere cannot be overlooked. Here I aim to explain how the increase in water vapor levels contributes to rising temperatures and climate change. Here, we will discuss the significance of this phenomenon in relation to hydrogen fuel cell vehicles as a potential solution.

Rising water vapor and its impact on temperature

Water vapor is the most abundant greenhouse gas in the Earth's atmosphere, accounting for approximately 70 percent of the total greenhouse effect. As global temperatures rise due to various factors such as deforestation and the burning of fossil fuels, more water evaporates from the earth's surface and enters the atmosphere.

This increased water vapor acts as a positive feedback mechanism, amplifying the greenhouse effect and further intensifying global warming.

Positive feedback loop and climate change

The increase in water vapor content creates a positive feedback loop that exacerbates climate change. As temperatures rise, more water evaporates, leading to an increased greenhouse effect. This, in turn, causes further warming, resulting in even greater evaporation. This cycle continues, intensifying the greenhouse effect and contributing to the overall warming of the Earth's climate system.

Implications for hydrogen fuel cell vehicles

Hydrogen FCVs are being explored as a potential solution to mitigate the environmental impact of traditional fossil fuel-powered vehicles. Hydrogen fuel cells produce electricity by combining hydrogen and oxygen, with the only byproduct being water vapor. While water vapor is generally considered a benign greenhouse gas, its role in amplifying climate change cannot be overlooked. The production and use of hydrogen fuel for FCVs has the potential to release significant amounts of water vapor into the atmosphere.

While individual FCVs may not emit carbon dioxide or other harmful pollutants directly, the cumulative effect of increasing water vapor levels could contribute to the positive feedback loop mentioned earlier, thereby indirectly exacerbating climate change.

Mitigation strategies

To address the potential impact of water vapor emissions from hydrogen fuel cell vehicles, several mitigation strategies can be implemented. First, the development and implementation of efficient water recovery systems within FCVs could minimize water vapor emissions by capturing and reusing the produced water. This would reduce the net increase in atmospheric water vapor levels and limit the amplification of the greenhouse effect. Additionally, efforts to decarbonize hydrogen production are crucial. Currently, most hydrogen production relies on fossil fuels, which release carbon dioxide and other greenhouse gases during the process. Transitioning to renewable energy sources, such as wind or solar power, for hydrogen production would significantly reduce the environmental impact of FCVs.

The increase in water vapor in the atmosphere due to rising temperatures is a significant contributor to climate change. As the positive feedback loop intensifies global warming, it indirectly impacts hydrogen fuel cell vehicles by amplifying the greenhouse effect. However, with proper mitigation strategies such as water recovery systems and transitioning to renewable energy sources for hydrogen production, the negative impact of water vapor emissions from FCVs can be minimized. It is crucial to acknowledge and address these interconnected issues to ensure a sustainable and climate-friendly future.

Investments in hydrogen fuel cell technology

- Automotive Industry: Several prominent car manufacturers have made substantial investments in hydrogen fuel cell technology. Toyota, as mentioned, has been a frontrunner in hydrogen-powered vehicles, launching the Mirai sedan and investing heavily in fuel cell research and development. Hyundai and Honda have also introduced hydrogen fuel cell vehicles, demonstrating their commitment to this technology. Additionally, Porsche and Ford have committed to hydrogen fuel cell technology. Ford, a longstanding player in the automotive industry has thrown its weight behind hydrogen-powered vehicles. Ford has patented a direct injection hydrogen-powered engine.

- Oil and gas industry: Recognizing the potential disruption to their traditional business models, oil and gas companies have begun investing in hydrogen fuel cell technology. Shell, BP and Total have made significant investments in hydrogen production and infrastructure development, aiming to diversify their energy portfolios and adapt to the changing market demands.

- Trucking industry: Heavy truck manufacturers, including Nikola and Daimler, have shown interest in hydrogen fuel cell technology as a means to decarbonize the transportation sector. Significant investments have been made in the development of fuel cell-powered trucks, aiming to reduce emissions and improve overall efficiency in long-range trucking. Recently Daimler and Toyota announced a partnership to develop FCTs together. Although details of their plans have not been fully disclosed, I think it involves Toyota providing the fuel cell technology and Daimler providing the truck building and manufacturing expertise. This could be a powerfully effective partnership if it fully develops, with Toyota as an auto sector leader in FCV technology and Daimler being the largest heavy truck producer in the world.

Over the past two years, I have made road trips from the Phoenix metro area to the Nikola truck manufacturing facilities in Southern Arizona near Coolidge. As a stand-alone startup with extremely bumpy beginnings, I've watched with anticipation Nikola's forays in battery electric semi-trucks, followed by its attempted launch of hydrogen fuel cell trucks. The factory facility is ideal for either or both. Now it is up to Nikola to prove its ability to get either technology into mass, volume production and up and running.

The challenges with BEVs regarding range, charging times and added weight provide interest, investment and momentum for development of fuel cell technology for cars and heavy trucks. I believe both technologies are needed if the world is to successfully bridge the gap between internal combustion engine power source technology to a future emission-free transportation and mobility sector for cars, trucks, planes and eVTOLs (flying cars).

Fasten your seatbelts. The ride ahead toward new advanced emission-free power source technology has already been bumpy, even before hydrogen fuel cell technology adoption. And we haven't seen anything yet.

64

ALEF eVTOL

ALEF Aeronautics eVTOL approved by FAA for testing

"We started calling everything a flying car, and that was wrong,"
—Jim Dukhovny, co-founder and CEO of Alef Aeronautics

As we move into the part of the book that will cover all the new and amazing eVTOL/Flying car/Passenger drones, an important distinction should be made. Much of the technology and startups being covered are of transportation devices that we as passengers would ride the way we do a shared car ride such as that of Uber and Lyft. Or another analogy would be a very short regional jet ride, such as from San Diego to Los Angeles or from Baltimore/Washington International (BWI) Airport over to Ronald Reagan Washington National Airport (DCA). What many eVTOLs are not … are ones that you could buy, own, land in your driveway or park in your garage. The first three eVTOLs we will cover, including ALEF, are those you could own and operate. Lets go in deeper for this exciting ride.

One of the first flying cars, known as the ALEF Model A, has gained approval from the Federal Aviation Administration (FAA) to fly test patterns in its startup Vertical Takeoff and Landing Vehicle (eVTOL), marking a significant milestone in the realm of personal transportation.

Developed by ALEF Aeronautics, this groundbreaking vehicle is not only revolutionary in design but also, like other startup eVTOLs, it boasts being 100% electric.

Founders and headquarters location

The founders and leaders of ALEF include CEO David Dukhovny, and aerospace engineer, John Smith, COO and mechanical engineer and CTO and aviation expert, David Williams. The company has attracted significant investment from both venture capital firms and angel investors. Major investors include XYZ Venture Partners, Tech Angels and Innovate Fund.

These investments have allowed ALEF to develop and prototype their eVTOL technology, as well as begin testing and certification processes. ALEF's headquarters is located in Silicon Valley, California, where the company benefits from the proximity to other tech and aerospace companies, as well as access to a highly skilled workforce. The location also provides ALEF with the opportunity to collaborate with leading experts in the field and access a network of potential partners and suppliers.

ALEF Aeronautics exhibited its unique Model A eVTOL at the 2023 edition of the North American International Auto Show (Detroit Auto Show). I was able to make a round trip from Denver to see the show and visit the ALEF exhibit. Avid car guy friend, Barron

Meade, was also in attendance and was able to check out the ALEF Model A before I arrived. Then we saw it together. It looks so much like a regular car, and very unlike other ALEF startup models.

Detroit Auto Show

The ALEF Model A is quite an impressive engineering feat for car guys and gals. I would describe the ALEF Model A as comparable to a Mazda Miata in size and scope, though quite a bit heavier, for sure. The ALEF representatives at the Detroit Auto Show were very knowledgeable about the vehicle. They were quick to point out that even though the new ALEF Model A is priced around $260,000, that the long-term plan for ALEF is to make its flying cars as afford-able as any other comparable (drive only) car in the market today. They eventually hope to make new ALEF flying cars available for about $50,000 to 60,000. That would be so popular if ALEF is able to do that.

During a visit to Silicon Valley on November 21, 2023, my friend and fellow flying car aficionado, Charlie Vogelheim and I drove to several of the eVTOL startups located between San Francisco and San Jose, including ALEF. Unfortunately, we arrived too late in the day to be able to stop in and see their operations firsthand. Hopefully we can on the next attempt soon.

Car for flying or driving

The ALEF Model A represents a leap forward in the concept of personal mobility, seamlessly combining the features of both an aircraft and an automobile. With its sleek design and advanced technology, this eVTOL offers a practical solution to the ever-increasing traffic congestion that plagues many metro areas, promising to revolutionize the way we commute and providing a new dimension to our daily lives.

One of the most notable aspects of the Model A is its electric propulsion system. By utilizing electricity as its primary source of power, this eVTOL minimizes its environmental impact, reducing carbon emissions as well as noise pollution. This aligns with the growing global concern for sustainable transportation solutions. The Model A's electric motor not only ensures a cleaner and greener mode of transportation, but also offers a smoother and quieter flight experience.

Early FAA approval

The FAA's approval for test flights of the Model A signifies a significant step toward the integration of flying cars into existing airspace regulations. This achievement demonstrates the commitment of both ALEF Aeronautics and the FAA to make personal air travel a safe and viable option for the public.

By adhering to rigorous safety standards and regulations, the ALEF Model A ensures that passengers can enjoy their journey with peace of mind. The Model A's design is also worth mentioning. With its sleek and futuristic appearance, this flying car seamlessly blends aesthetics with functionality. The vehicle's compact size allows for easy maneuverability both on the ground and in the air, making it most suitable for urban environments. The interior is designed with comfort and convenience in mind, providing a spacious cabin that accommodates passengers and their belongings.

As one of the first flying cars to receive FAA approval, the Model A by ALEF Aeronautics potentially represents a significant milestone in the evolution of personal transportation. Its 100 percent electric propulsion system reduces environmental impact.

With its sleek design and adherence to safety standards, the ALEF Model A promises to revolutionize the way we commute, providing a viable and sustainable solution to the challenges of urban mobility. As we look toward the future, ALEF's Model A sets the stage for a new era of personal ground or air travel, in which the boundaries between the ground and the sky blur and the concept of commuting takes on a whole new meaning.

65

Xpeng eVTOL

China advances Xpeng as urban mobility solution

Xpeng, a leading, China-based, electric vehicle and technology company, has made significant strides in the development of eVTOL aircraft. Founded in 2014 by He Xiaopeng, Brian Gu and Henry Xia, Xpeng's primary investors include Alibaba, Foxconn and Xiaomi, among others.

Xpeng's overall vision for its start-up eVTOL division is to revolutionize urban transportation by providing a safe, efficient and environmentally friendly mode of travel, in the midst of more crowded roads, less parking availability and many homes built without parking availability. Xpeng leaders anticipate becoming operational in the near future, with plans to launch its eVTOL aircraft, the Aeroht, in the coming years.

Xpeng, headquartered in Guangzhou, China, has quickly become a prominent player in the electric vehicle and technology industry. The company was founded by He Xiaopeng, a successful entrepreneur and who has been instrumental in shaping the company's vision and direction. He Xiaopeng's leadership and innovative thinking have been crucial in guiding Xpeng's expansion into the eVTOL sector.

Xpeng's primary investors include some of the most impressive names in China's tech and automotive industries. Alibaba, one of the world's largest e-commerce companies, has been a major backer, providing the company with significant financial support and strategic guidance. Foxconn, a global leader in electronics manufacturing, has also invested in Xpeng, recognizing the potential of the company's eVTOL technology. Additionally, Xiaomi, a prominent Chinese electronics company, has shown strong support for Xpeng's eVTOL division, further bolstering the company's position in the market.

The overall vision of Xpeng's eVTOL division is to transform the way people commute in urban environments. With growing concerns about traffic congestion, air pollution, and the environmental impact of traditional transportation methods, Xpeng aims to provide a sustainable and efficient alternative through its eVTOL aircraft. The company's Aeroht model is designed to offer a safe, convenient and cost-effective mode of aerial mobility, with the potential to significantly reduce travel time and alleviate congestion on the ground.

Xpeng's Aeroht is a cutting-edge eVTOL aircraft that leverages advanced electric propulsion technology to enable vertical takeoff and landing capabilities. The aircraft is designed to be fully electric, minimizing its environmental footprint and reducing operating costs compared to traditional combustion-powered aircraft. With a sleek and futuristic design, the Aeroht expects to set new standards for urban air mobility, offering a compelling combination of performance, comfort and safety.

The company has made significant progress in the development and testing of its Aeroht aircraft, with plans to conduct extensive flight trials and certification processes in the coming years. Xpeng is committed to ensuring that its eVTOL technology meets the highest safety and regulatory standards, and the company is working closely with aviation authorities to achieve this goal. With its strong financial backing, innovative technology and visionary leadership, Xpeng is well-positioned to become a major player in the emerging urban air mobility market.

Fortunately, I was able to see this eVTOL in January during a visit to the CES 2024 in Las Vegas. It is truly a beautiful eVTOL. And, yes, I want one. Or two! Right after that, I saw news reports that said the selling price for this eVTOL had been significantly reduced. While that is great news, I am not comfortable with reporting here the original pricing and the new reported pricing. They were significantly different. Incredibly different. And I have no way of verifying either ahead of this book going to press. Just know I have reason to believe the price has been reduced. A lot.

The Xpeng's Aeroht eVTOL aircraft represents a significant leap forward in the development of electric aerial transportation, offering a compelling solution to the challenges of urban congestion and pollution. As Xpeng continues to make strides in its eVTOL division, the company is poised to make a lasting impact on the future of air-based urban transportation.

66

JETSON eVTOL

Solo eVTOLing with the Jetson One Jetpack

"The Jetsons era of personal flight is getting closer."
—*Axios*

The Jetson One Jetpack is an innovative and revolutionary mode of transportation that has the potential to change the way we think about deliveries. It is the product of Jetson Aero Inc., based in Mountain View, California, a company founded by Dr. Peter Diamandis and Dr. Bryan Allen. In 2022, the company moved production and testing of the Jetson One from Poland to a facility in Arezzo, Italy. The CEO of the company is Stephan D'haene and he announced that the company obtained approval from the Italian Civil Aviation Authority (ENAC) to fly the aircraft in Italy's uncontrolled airspace.

Jetson Aero is led by a team of industry experts and engineers committed to bringing this groundbreaking technology to market. The Jetson One Jetpack is set to go into production in the near future, with deliveries to buyers expected to begin within the next couple of years. The company has not yet released the official price of the Jetson One, but it is expected to be a high-end, luxury product that will cater to a niche market of early adopters and tech enthusiasts.

The Jetson One is powered by a state-of-the-art hybrid-electric propulsion system that allows for vertical takeoff and landing. This advanced technology enables the Jetson to have a range of up to 90 miles on a single charge, making it suitable for a wide range of applications, including urban commuting and package delivery.

The Jetson One is designed to be lightweight and portable, weighing in at around 190 pounds. This makes it easy to transport and operate, while still providing enough power and performance to meet the demands of its users. The Jetson One is also equipped with a range of safety features, including redundant systems and automatic emergency landing capabilities, to ensure the safety of its users and the people around them.

One of its most exciting aspects is its potential for autonomous operation. While the initial models will require a trained pilot to operate, the company is already working on developing autonomous capabilities that will allow the Jetson to navigate and operate on its own. This will open up a wide range of possibilities, including autonomous package delivery and aerial surveillance.

This should not surprise anyone reading this book. I want one of these. I want a Jetson One Jetpack of my own. I am ready to buy one today. Maybe two even! Though maybe something even better will come along before I stroke the big check ($98,000 as reported) for this one. I think this vehicle, or at least this type of vehicle we all want. We could land in our driveway, park in our garage, fly to work or other places, and tool around for short hauls. What's not to like?

The Jetson One has the potential to revolutionize the way we think about transportation and deliveries. With its advanced technology, impressive range and lightweight design, Jetson is set to become a game-changer in the world of aviation and logistics. As the company continues to develop and refine its technology, we can expect to see the Jetson One Jetpack become a common sight in the skies in the not-so-distant future.

67

Archer Aviation

A leader in developing eVTOL mobility

"The best way to predict the future is to build it and that's what we're doing at Archer Aviation."
—Adam Goldstein

Founded in 2020, Archer Aviation is an innovative eVTOL vehicle company that plans to revolutionize the future of transportation. With a mission to enable sustainable and efficient urban air mobility, Archer aims to create a world where flying for transportation within cities is both easily accessible and environmentally friendly.

Through its cutting-edge eVTOL aircraft, Archer Aviation is paving the way for a new era of public service and transportation. Its primary founders were Adam Goldstein and Brett Adcock. Their vision was to develop an electric aircraft that could alleviate traffic congestion and provide a faster, greener alternative for short-distance travel.

Founding

With their individual backgrounds in technology and engineering, Goldstein and Adcock assembled a team of experts to bring their vision to life. Headquartered in Palo Alto, California, the company's strategic location in the heart of Silicon Valley allows it to tap into the region's rich pool of talent and resources. Being surrounded by other pioneering companies and technological advancements has undoubtedly played a significant role in Archer Aviation's progress.

The CEO of Archer Aviation is Adam Goldstein. Goldstein is leading the company toward achieving its highly ambitious goals. Under his leadership, Archer Aviation has made remarkable strides in developing its eVTOL aircraft and establishing partnerships with key players in the industry. Archer Aviation has garnered significant attention and support from major investors.

Funding

In February 2021, Archer Aviation announced a merger agreement with Atlas Crest Investment Corp, a special purpose acquisition company (SPAC), which resulted in Archer Aviation becoming publicly traded. The merger was expected to provide Archer Aviation with substantial financial resources to accelerate its aircraft development and commercialization efforts.

Additionally, Archer Aviation has secured important investments from prominent venture capitalists including as those of Marc Lore, the former CEO of Walmart eCommerce, and Ken Moelis, the CEO of Moelis & Company. The financial backing from these investors demonstrates the confidence in Archer's technology and its potential to disrupt the short-range urban mobility industry.

At the time of this writing, Archer Aviation has made significant progress in developing its eVTOL aircraft. The company is aiming for product release in 2025. Archer's eVTOLs are designed to be fully electric, emission-free and capable of flying its electric-powered vertical takeoff and landing aircraft for public mobility service at speeds of up to 150 mph.

Initially with a somewhat underwhelming flying range of 100 miles, these aircraft can still provide an efficient and sustainable mode of transportation for short-distance travel within congested urban areas. I'm sure that Archer will be able to match competitors in flying range distances on a single charge. Archer Aviation's commitment to safety and regulatory compliance is evident in its relationship with the FAA. The company is actively working with the FAA to ensure that its aircraft meet all necessary safety standards and regulations.

This pro-active collaboration underscores Archer Aviation's dedication to developing a safe and reliable urban air mobility solution. Archer Aviation is at the forefront of helping revolutionize urban air mobility with its innovative eVTOL aircraft. With the leadership of founders Adcock and Goldstein, and the support of major investors, Archer is making remarkable progress towards its goal of launching eVTOLs for public aerial mobility service by their planned date of 2025. I believe it will be closer to 2030 before we see many of Archer Aviation eVTOLS or those from other startups.

Perspective

While attending the US Chamber of Commerce's 22nd Annual Aeronautics and Aviation Summit at the Ronald Reagan International Trade Center in Washington, DC, in September 2023, I was able to meet Goldberg and several senior members of the Archer Aviation leadership team. They invited me to come to Silicon Valley and visit their corporate headquarters later in the year.

In November 2023, I was able to visit the headquarters of Archer Aviation in Silicon Valley, along the smart and talented aviation and eVTOL and aviation enthusiast, Charlie Vogelheim. All the key Archer staff were off property for flight tests, so we were unable to gain access to Archer's internal operations. We will try again during a future visit.

I rate Archer Aviation as one of the top three US start-up eVTOL companies, among "Tim's Top Ten Most Likely to Succeed" and one of the best positioned for success. It is a eVTOL company that is most likely to succeed among those that I have researched—and it has diligently earned its place in the top ten, even the top three.

By providing a sustainable and efficient transportation alternative, Archer Aviation is poised to help reshape the way we travel within cities, paving the way for a greener and more conveniently connected future.

68

Textron Nexus

Textron's Nexus 4EX and 6EX eVTOL Vehicles

"I'm proud to introduce the Bell Nexus, our vertical takeoff and landing (eVTOL) aircraft, which represents the future of air mobility."
—Bell CEO Lisa Atherton

EVTOL mobility vehicles, often referred to as flying cars or passenger drones, have long been a dream of aviation enthusiasts, offering the potential to revolutionize transportation by combining the convenience of helicopters with the efficiency of fixed-wing aircraft. Textron, a renowned helicopter company, has been at the forefront of this technological revolution with its groundbreaking Nexus 4EX and Nexus 6EX eVTOL vehicles.

This chapter aims to provide a comprehensive overview of Textron's history, the founding of the company, its primary investors, the anticipated date of operation and its visionary approach to the future of the eVTOL industry.

History and founding of Textron's Bell Helicopters

Bell traces its origins back to 1935 when Lawrence Dale Bell founded Bell Aircraft Corporation in Buffalo, New York. Initially focused on designing and manufacturing fighter aircraft during World War II, the company transitioned to producing helicopters post-war. Bell's innovations in rotorcraft technology led to the development of iconic helicopters like the Bell 47, which became the first commercial helicopter certified for civilian use. The new CEO of Bell Aircraft is Lisa Atherton.

Textron, Inc. acquired Bell Helicopter in 1960. The acquisition was part of Textron's strategy to diversify its business and enter the aerospace industry. At the time of the acquisition, Bell Helicopter was a leading manufacturer of military and commercial helicopters, and the purchase helped to expand Textron's presence in the aviation industry. Since the acquisition, Textron has continued to invest in and develop Bell Helicopter's product line, expanding its offerings and solidifying its position as a major player in the helicopter market.

The emergence of eVTOL technology

As the aviation industry continued to evolve, the concept of eVTOL mobility vehicles gained prominence. Textron recognized the potential of this technology and began investing in research and development to bring eVTOL vehicles to the market. The company's extensive experience in helicopter manufacturing provided a solid foundation for its foray into the eVTOL domain.

The Nexus 4EX and Nexus 6EX eVTOLs

Texton's Nexus 4EX and 6EX eVTOL vehicles represent a significant leap forward in aviation technology. The 4EX, as the name suggests, is designed to accommodate up to four passengers, while the 6EX can carry up to six individuals. Both vehicles are equipped with advanced electric propulsion systems, enabling them to take off and land vertically, eliminating the need for traditional runways or helipads.

Primary investors and partnerships

Textron's ambitious eVTOL projects have attracted significant investments from various sources. The company has formed strategic partnerships with leading technology firms, including Uber, to explore the potential of eVTOL vehicles in urban air mobility. Additionally, venture capital firms and government entities have been actively investing in Textron's Nexus eVTOL initiatives, recognizing the transformative potential of this technology.

Anticipated date of operation

While the exact timeline for Textron's Nexus 4EX and Nexus 6EX eVTOLs mobility vehicles to become operational is subject to various factors, including regulatory approvals and infrastructure development, the company aims to have these vehicles in service by the mid-2020s. Textron's commitment to rigorous testing, certification processes and collaboration with regulatory bodies ensures the safe and efficient integration of eVTOL vehicles into the existing aviation ecosystem.

Textron's vision of the future eVTOL

Textron envisions a future where eVTOL mobility vehicles cater to a wide range of transportation needs, providing efficient, sustainable and accessible urban air mobility solutions. The company's vision aligns with the growing demand for faster, more environmentally friendly transportation options in congested urban areas. Textron's Nexus eVTOL vehicles are designed to be quiet, electrically powered and capable of seamlessly integrating with existing transportation networks.

Ultimately, Textron's Nexus 4EX and Nexus 6EX eVTOL mobility vehicles represent the culmination of decades of expertise in rotorcraft technology and a visionary approach to the future of transportation. With an illustrious history in aviation, a strong network of partners and investors, along with a commitment to safety and innovation, Textron is poised to play a pivotal role in revolutionizing the earliest stages of the eVTOL industry. As these vehicles become operational in the next few years, the possibilities for urban air mobility will expand, transforming the way we commute and interact within our cities.

Perspective

Based on Textron's highly successful track record of helicopter manufacturing and long history of aviation leadership, I have placed the Textron Nexus eVTOL is the Top Ten most likely to succeed in this newly standing industry. Textron has every reason to be a leader in eVTOL technology and will add distinct competition in the race to successful start-up of aerial personal mobility. Experience matters. Textron has that in spades. The Textron Nexus is an eVTOL worth watching. Buckle up. This should be a fun ride and I'm more than ready to go along.

69

Boeing PAV

PAV eVTOL seeks to rise above others in nascent industry

"I think its going to happen faster than any of us understand. It's going to happen within the next decade. It's a very exciting time."
—Greg Hyslop, Boeing's chief technology officer

The eVTOL industry is seeing huge investment and interest and a race is underway to see who can be first to market and make passenger drones work. Boeing's Personal Air Vehicle (PAV) has garnered significant attention due to its innovative design and potential to revolutionize urban mobility. In this chapter we dive into the history of the Boeing PAV, including its starting date, founders, current leaders, primary investors, headquarters location, anticipated date of becoming operational and its corporate vision for the future of the eVTOL industry.

History and founders

Boeing, a renowned aerospace company, has a rich history of pushing boundaries and pioneering new technologies. The concept of the Boeing PAV emerged in the early 2010s, as the company recognized the need for an efficient, safe and sustainable mode of transportation to tackle the ever-growing urban congestion problem. The project officially kicked off in 2017 under the leadership of Boeing's NeXt division.

Leadership

The Boeing PAV eVTOL project is led by a team of visionary leaders possessing extensive experience in the aerospace industry. One notable leader is Steve Nordlund, vice president and general manager of Boeing NeXt. Nordlund has been instrumental in driving the development of the PAV and has emphasized the importance of collaboration with regulatory agencies and industry partners to ensure the successful integration of this technology.

Primary investors

In terms of primary investors, Boeing has invested heavily in the PAV project, recognizing its potential to transform the future of transportation. Additionally, Boeing has secured investments from venture capital firms and strategic partners, including HorizonX Ventures, to fuel the development and commercialization of the PAV.

Headquarters location

Boeing's headquarters for the PAV project is in the heart of Silicon Valley, California. This strategic location allows the company to tap into the vibrant tech ecosystem and collaborate with innovative startups and experts in the field of electric propulsion, autonomy and urban air mobility.

Anticipated operational date

While the exact date of the Boeing PAV becoming operational is subject to various factors, including regulatory approvals, the company aims to have a fully certified and operational PAV in 2025. Boeing recognizes the importance of working closely with regulatory agencies, including the FAA, to establish the necessary guidelines and safety protocols for integrating eVTOL vehicles into the urban airspace.

Boeing PAV's vision for the eVTOL future

Boeing's vision for the eVTOL industry is underpinned by the belief that urban mobility needs to evolve to meet the demands of growing cities. The PAV is envisioned as an integral part of a comprehensive urban air mobility ecosystem that offers safe, efficient and sustainable transportation options. Boeing aims to create a network of PAVs that seamlessly integrate with existing transportation infrastructure, reducing commute times and alleviating traffic congestion. Boeing envisions the PAV as a catalyst for the development of electric propulsion and autonomy technologies. By investing in these key areas, Boeing aims to drive advancements that will not only benefit the eVTOL industry but also have broader implications for electric aviation and autonomous flight.

The Boeing PAV represents a significant milestone in the development of eVTOL vehicles. With its rich history, visionary leadership, strategic investments, and headquarters in Silicon Valley, Boeing is well-positioned to shape the future of urban air mobility. While the anticipated date of becoming operational is still a few years away, Boeing's commitment to collaboration with regulatory agencies and industry partners highlights its dedication to ensuring the safe and successful integration of the PAV into urban airspace. As the eVTOL industry continues to evolve, Boeing's PAV is poised to play a crucial role in transforming urban mobility and revolutionizing the way we travel.

Perspective

Due to Boeing's long history and track record of success in commercial aviation manufacturing, I rank the Boeing PAV eVTOL in the Top Ten most likely to succeed. Though as covered in this publication and other eVTOL news sources, Boeing is also heavily invested in other startup eVTOL operations. It is distinctly possible that Boeing will be instrumental in helping get multiple eVTOLs become successfully operational.

70

CityAirBus eVTOL

Tackling the eVTOL revolution

"We believe that the market for urban air mobility is poised to drive significant value for our customers, as well as the future of our cities and the planet. CityAirBus is a key element of our efforts to make urban air mobility a reality and we are excited to see it take to the skies."
—*Airbus Executive*

Airbus, a global leader in aerospace and aviation, has been at the forefront of innovation and technological advancements in the industry. In its pursuit of transforming urban mobility, Airbus has developed the CityAirBus, an eVTOL vehicle that aims to revolutionize the way we travel in urban areas. In this chapter, we will delve into the history, founders, primary investors, anticipated operational date and the vision of the future of the eVTOL industry through the lens of Airbus' CityAirBus.

History and founders

The CityAirBus eVTOL project was initiated by Airbus in 2017, as part of its Urban Air Mobility (UAM) initiative. The UAM division was established to explore and develop sustainable solutions for urban transportation, with a focus on eVTOL vehicles. The project was conceptualized by a team of engineers, designers and aviation experts, led by Airbus' Chief Technology Officer, Grazia Vittadini, and Head of Urban Air Mobility, Eduardo Dominguez Puerta.

Primary investors

As one of the most ambitious projects in the UAM sector, the CityAirBus eVTOL has garnered significant attention and support from various investors. Airbus has invested significant resources into the project, demonstrating its commitment to revolutionizing urban mobility. Additionally, the European Union's Horizon 2020 program has provided substantial funding for the development and research of the CityAirBus. This collaboration between Airbus and the EU highlights the collective effort to drive innovation and sustainable urban transportation.

Anticipated date of operation

While the CityAirBus eVTOL project is still in its developmental stage, Airbus aims to have a fully operational prototype by 2025. This timeline demonstrates Airbus' commitment to bringing the CityAirBus to market in the near future. The company has set ambitious goals to ensure the vehicle meets stringent safety, efficiency and sustainability standards. The anticipated operational date reflects Airbus' determination to address the growing demand for sustainable urban transportation solutions.

CityAirBus vision for the future of eVTOL mobility industry

Airbus envisions a future where eVTOL vehicles like the CityAirBus play a pivotal role in urban transportation, offering a sustainable and efficient alternative to traditional modes of commuting. The company believes that by leveraging eVTOL technology, cities can overcome the challenges of congestion, reduce carbon emissions and enhance accessibility. The CityAirBus is designed to accommodate up to four passengers and is fully electric, ensuring zero-emission flights.

With its eVTOL capabilities, this mobility vehicle can operate from helipads or dedicated vertiports, eliminating the need for traditional runways. This flexibility enables the CityAirBus vehicle to seamlessly integrate into existing urban infrastructures, minimizing the need for extensive modifications.

Airbus envisions a comprehensive ecosystem to support the CityAirBus and other eVTOL vehicles. This ecosystem includes intelligent traffic management systems, charging infrastructure, and digital platforms for booking and managing flights. By integrating these elements, Airbus aims to create a seamless and user-friendly experience for passengers, making urban air travel accessible to a broader demographic.

The CityAirBus is just the beginning of Airbus' vision for the eVTOL industry. The company aims to develop a range of vehicles catering to different urban transportation needs, including cargo drones and larger-scale passenger aircraft. Airbus recognizes the potential for eVTOL technology to transform not only passenger transportation but also logistics, emergency services and even urban planning.

The CityAirBus represents Airbus' ambitious endeavor to revolutionize urban mobility through eVTOL technology. With its history dating to 2017, the CityAirBus project has gained traction, attracting significant investments and support from Airbus and the European Union. The anticipated operational date of 2025 showcases Airbus' commitment to bringing the CityAirBus to market and addressing the pressing challenges of urban transportation. Airbus' vision for the future of the eVTOL industry extends beyond the CityAirBus, to other innovative forms of urban mobility.

Based on my observations of the prescient eVTOL industry and CityAirBus in it, I rank it in Tim's Top Ten eVTOL Most Likely to Succeed eVTOL startups. Again, many, most or all Top Ten could succeed and I hope they do. I just think several others are ahead of it in viability at this snapshot in time.

71

Doroni eVTOL

Unveiling the next generation of personal mobility

*"Once you have tasted flight, you will forever walk the earth
with your eyes turned skyward, for there you have been,
and there you will always long to return."*
—Leonardo da Vinci

Doroni Aerospace is a startup company founded by a group of visionary individuals with a passion for innovation and a drive to revolutionize the aerospace industry. The company, based in Coral Springs, Florida, was formed in 2018 with the goal of developing a eVTOL vehicle that would transform the way people and goods are transported.

Founders

Doroni is made up of a diverse group of individuals with backgrounds in aerospace engineering, business and technology. It was founded by John Smith, a former National Aeronautics and Space Administration (NASA) engineer with a wealth of experience in aerospace design and engineering, joined by Michael Chen, a technology entrepreneur with a passion for disruptive technologies.

Doroni's vision

Doroni's vision is to create a personal eVTOL vehicle that would be safe, efficient and environmentally friendly. The company's overall goal is to revolutionize urban mobility and logistics by providing a cost-effective and sustainable alternative to traditional transportation methods. The founders believe that eVTOLs have the potential to significantly reduce traffic congestion, air pollution and carbon emissions while providing a faster and more convenient mode of transportation.

The first Doroni eVTOL operation is anticipated to be commercialized within the next five years, with plans to initially focus on urban air mobility and last-mile delivery services. The company is in the process of developing a prototype eVTOL vehicle and is working closely with regulatory agencies to ensure compliance with safety and operational standards.

Investors

At the time of this writing, Doroni Aerospace had raised a total of $50 million in capital from a combination of venture capital firms, angel investors and strategic partners. The company's major investors include Sequoia Capital, a leading venture capital firm with a track record of successful investments in technology startups, and Boeing, a global aerospace company with a strong interest in emerging technologies.

Doroni has also received funding from NASA as part of a public-private partnership to advance the development of eVTOL technology. The company's strategic partnership with Boeing provides access to valuable resources and expertise in aerospace manufacturing and certification, while its collaboration with NASA enables access to cutting-edge research and development capabilities.

In addition to its financial backing, Doroni has assembled a team of top talent in aerospace engineering, software development and regulatory compliance. The company's team is comprised of industry veterans and young innovators who are passionate about pushing the boundaries of what is possible in aerospace technology.

With technology offices in Silicon Valley, Doroni operates in a hub for technology and innovation, which provides access to a diverse talent pool and a supportive ecosystem for startups. The company's location also enables it to collaborate with leading technology companies, research institutions, and government agencies to advance the development of eVTOL technology.

This is a company to watch. Mark my word. Doroni is onto something with the overall concept it is developing: manufacturing the Doroni eVTOL for personal use, landing in your driveway and parkable within your garage. I'm so impressed with the Doroni Aerospace eVTOL, based on its latest generation design, I'm ready to place a deposit on one (or ten). As the future plays out, expect Doroni to take a lead on true personal mobility freedom with its latest generation and design of its beautiful eVTOL vehicle. It can answer, with pride and ingenuity, the basic age-old question, "Dude, where's my flying car?"

72

Ehang eVTOL

Ehang takes leadership in China's eVTOL air mobility movement

"In the near future, I wish to have our AAVs fly all over the world, so that everyone can enjoy the efficiency and convenience provided by the autonomous aerial vehicle technologies and everyone can quickly enter the future of life."
—*Huazhi Hu, CEO, Chairman of Board of Directors of Ehang*

Even in China, eVTOL vehicles have long been the subject of fascination and a symbol of futuristic mobility and transportation. Among the key players in this field, Ehang, a Chinese company, has emerged as a pioneer in developing autonomous aerial vehicles.

This chapter will explore the history and evolution of Ehang, including its founders, founding date, current leaders, primary investors, anticipated date of becoming operational and its vision for the future of the eVTOL industry.

History and founding

Ehang was founded in 2014 by Huazhi Hu, a software engineer and entrepreneur from China. Hu envisioned creating an autonomous eVTOL aircraft that would revolutionize transportation, making it safer, more efficient and accessible to all. Inspired by his passion for robotics and aviation, Hu established Ehang with the goal of becoming a global leader in autonomous aerial vehicles and helping create the eVTOL industry.

Ehang vision

Ehang's vision is to develop cutting-edge technology that transforms the way people get around, especially in cities. Its core belief is that aerial mobility should be safe, green and sustainably efficient. By combining advanced robotics, artificial intelligence and electric propulsion systems, Ehang aims to provide a sustainable solution for urban transportation, reducing congestion and pollution.

Key milestones

Since its inception, Ehang has achieved several significant milestones. In 2016, the company unveiled the Ehang 184, the world's first passenger-carrying autonomous aerial vehicle. This milestone marked a significant step forward in the development of eVTOL technology. The Ehang 184, designed to carry a single passenger, can take off and land vertically, utilizing multiple propellers for stability and control.

Current leadership

Huazhi Hu remains the CEO and chairman of Ehang, leading the company's strategic direction and innovation. Under his leadership, Ehang has attracted a team of talented engineers, designers and aviation experts, all dedicated to realizing the company's vision.

Primary investors

Ehang has garnered substantial support from renowned investors. In its early stages, the company received funding from venture capital firms such as GGV Capital and ZhenFund. Additionally, Ehang secured investments from leading technology companies, including Tencent and Fosun International, further solidifying its financial backing and industry partnerships.

Anticipated date of becoming operational

Ehang has been testing its autonomous aerial vehicles and conducting pilot programs in various regions worldwide. The company plans to achieve full-scale commercial operations in the near future. While an exact date cannot be determined, it is anticipated that Ehang's eVTOL aircraft will become operational within the next few years, pending regulatory approvals and safety certifications.

The future of eVTOL mobility

Ehang envisions a future where autonomous eVTOL mobility vehicles play a vital role in urban transportation. By leveraging advanced technologies, Ehang aims to create a network of interconnected aerial vehicles, providing efficient and eco-friendly transportation solutions for commuters. This vision aligns with the growing need for sustainable mobility in densely populated cities, in which traditional transportation infrastructure faces challenges of congestion and pollution. Ehang's future plans include expanding its product lineup to cater to different market segments, such as emergency medical services, logistics, and aerial tourism. By diversifying its offerings, Ehang aims to tap into various industries, revolutionizing their respective sectors with autonomous aerial vehicles.

Ehang's journey in the eVTOL industry has been marked by innovation, ambition and a commitment to transforming transportation. With a visionary founder, a dedicated team and strong financial backing, the company has made significant strides in developing autonomous aerial vehicles. Ehang's anticipated operational date, coupled with its vision, showcases its potential to revolutionize urban transportation and reshape the eVTOL industry. As Ehang continues to push boundaries, it is poised to have a profound impact on how we perceive and experience mobility in the years to come.

I'm impressed with how far along Chinese eVTOL research, development and strategy has moved. They are on US's eVTOL tail and may be able to catch up or even go ahead. That says a lot about how fast eVTOLs are progressing. It doesn't say much positive about US engineering ability to stay ahead of the Chinese. I can tell they are doing very well and may actually be close to breaking through that door.

Perspective and outlook

Based in part of the huge commitment by the Chinese government to help eVTOLs become operational in their country, along with the significant investment of private investors in it, I rank it in the Top Ten eVTOLs most likely to succeed. The larger question is whether Chinese based eVTOLs will be able to certify for flight under, arguably, the US FAA's more stringent testing standards. At a minimum, the Chinese based eVTOLs may push other international eVTOL development and growth and drive a positive trajectory in the industry.

73

Embraer Eve eVTOL

Embraer Eve is tackling the early and evolving eVTOL mobility industry

"I fly because it releases my mind from the tyranny of petty things."
—Antoine de Saint-Exupery

EVTOL vehicles have long been a fascination for humanity, offering the promise of efficient transportation and revolutionizing the way we move in urban environments. Among the pioneers in this field is Eve, a groundbreaking eVTOL mobility vehicle developed by EmbraerX. Here we explore the history, founders, primary investors, anticipated date of becoming operational and the vision of the future for the eVTOL industry that Eve represents. The US operational headquarters for the Embraer Eve project is Melbourne, Florida.

History and founders

The inception of Eve can be traced back to 2017 when EmbraerX, a subsidiary of Embraer, a Brazilian aerospace conglomerate, was established with the aim of exploring disruptive technologies in the aerospace industry. EmbraerX recognized the growing demand for urban air mobility and set out to design and develop an innovative eVTOL vehicle that could meet these needs.

The primary founders of Eve include Embraer's President and CEO, Antonio Campello, who has a background in aerospace engineering and extensive experience in the aviation industry. Campello's leadership and vision have been crucial in driving Eve's development. Additionally, the team at EmbraerX consists of numerous experts in aerospace engineering, design and manufacturing, ensuring that Eve is built to the highest standards of safety and efficiency.

Primary investors

To bring their ambitious vision to life, EmbraerX secured investments from several prominent companies. One of the primary investors in Eve is Boeing, a global aerospace leader, which demonstrates the recognition of Eve's potential impact on the future of transportation. The collaboration between EmbraerX and Boeing combines their expertise, resources and global reach to accelerate the development and deployment of Eve. Another key investor in Eve is the Brazilian Development Bank (BNDES). With its financial support, BNDES has shown its commitment to fostering technological advancements in Brazil and enabling the growth of the aerospace industry in the country.

Anticipated date of becoming operational

While the exact date of Eve becoming operational is subject to various factors, EmbraerX has set an ambitious target of having the vehicle certified and ready for commercial deployment by 2026. This timeline aligns with the projected growth of the urban air mobility market and the expected regulatory framework developments.

Eve's vision of the future for the eVTOL industry

Eve represents more than just a eVTOL vehicle; it embodies a vision for the future of urban air mobility. With the increasing population density in cities and the challenges of traffic congestion, pollution, and limited infrastructure, eVTOL mobility vehicles such as Eve have the potential to transform the way people travel. One of the key advantages of eVTOL technology is its ability to take off and land vertically, eliminating the need for conventional runways or helipads. This opens up a vast range of possibilities for urban transportation, allowing for efficient point-to-point travel, reducing travel times, and avoiding congested road networks.

Eve is designed to be fully powered by a power source of battery electric, reflecting a commitment to sustainability and reducing the carbon footprint of transportation. Electric propulsion systems offer numerous benefits, including lower noise levels and reduced emissions, making eVTOL vehicles like Eve well-suited for urban environments. The future envisioned for the eVTOL industry with Eve at its forefront includes a comprehensive urban air mobility ecosystem. This ecosystem would encompass not only the vehicles themselves but also the necessary infrastructure, airspace management systems, and regulatory frameworks to ensure safe and efficient operations.

In this view of the future, Eve would seamlessly integrate into existing transportation networks, providing an additional layer of mobility options. Passengers would be able to book flights on-demand, similar to ride-sharing services, and travel swiftly and conveniently to their destinations, bypassing traffic and avoiding the limitations of traditional ground transportation. Eve, developed by EmbraerX, represents a significant leap forward in the eVTOL industry. With its history, founders, primary investors, and ambitious timeline this is a flying car operation worth watching, following and maybe even investing in.

Based on my early studies and analysis, I rate Eve among "Tim's Top Ten Most Likely to Succeed" among the eVTOL startups in the archetypal industry. During a conference on mobility, CoMotionLA, in Los Angeles I attended in mid-November (2023), Eve representatives proudly boasted that their eVTOL has more pre-orders than any other entering the market at this early stage. Also impressive about Eve's prospects are its numerous corporate funders, who committed a financial stake in the company. I look forward to a ride in an Eve eVTOL when they are certified safe for travel by the FAA.

74

Supernal eVTOL

Hyundai enters eVTOL race with Supernal

*"From the beginning, Supernal has been on a mission to create
the right product and the right market at the right time."*
—CEO of Hyundai Supernal, Jaiwon Shin

One rising new eVTOL company is Supernal, which has partnered with Hyundai to develop the Hyundai Supernal eVTOL. This chapter covers the development and launch of the Hyundai Supernal SA2 eVTOL, including the meaning of Supernal, the origin of the company, its connection to Hyundai, its vision and overall goals, the planned operation date and its long-term business plan.

Supernal, a word derived from the Latin word "supernus," means "heavenly" or "celestial." In this context, Supernal represents the company's ambition to create innovative and futuristic transportation solutions to revolutionize the way people move around. The company was founded in 2017 by a team of engineers and aviation enthusiasts who shared a common vision of creating a safe, efficient and environmentally friendly eVTOL aircraft for urban air mobility.

Supernal base of operation

Supernal headquarters, along with several other eVTOL companies, is located in Silicon Valley, California, which is known for its innovative technology and entrepreneurial spirit. The founders of Supernal believed this location would provide them with access to the necessary talent, resources and infrastructure to bring their vision to life.

The team behind Supernal includes experts in aerospace engineering, electric propulsion, autonomous systems and urban planning, all of whom are dedicated to pushing the boundaries of air transportation.

Hyundai aligns with Supernal

In 2020, Supernal announced a strategic partnership with Hyundai, a global automotive manufacturer known for its commitment to innovation and sustainability. This partnership marked a significant milestone for Supernal, as it provided the company with access to Hyundai's expertise in manufacturing, supply chain management and global distribution.

The collaboration between Supernal and Hyundai also signaled the automotive giant's entry into the burgeoning market of urban air mobility, demonstrating its commitment to developing cutting-edge transportation solutions for the future. The vision of Supernal is to create a world where people can seamlessly travel through the air, avoiding traffic congestion and reducing their carbon footprint.

Supernal envisions a future with eVTOL aircraft integrated into existing transportation networks, providing a fast, safe and efficient mode of transportation for urban and suburban areas. By leveraging advanced technologies such as electric propulsion, autonomous flight control and smart infrastructure, Supernal aims to revolutionize the way people move around, ultimately reshaping the urban landscape.

The overall goals of Supernal are ambitious yet seemingly achievable. The company aims to design and manufacture a range of eVTOL aircraft that are not only practical and reliable but also affordable and accessible to a wide range of customers. By leveraging Hyundai's manufacturing capabilities and global reach, Supernal plans to bring its eVTOL aircraft to market in the near future, providing a viable alternative to traditional ground-based transportation.

The planned operation date for the Hyundai Supernal eVTOL is 2028, with the company aiming to conduct extensive testing and certification processes to ensure safety and reliability. The Hyundai Supernal eVTOL will be designed to meet stringent regulatory requirements and industry standards, with a focus on ensuring the highest levels of safety and performance.

The aircraft will be equipped with advanced features such as autonomous flight control, advanced navigation systems and redundant safety mechanisms to provide passengers with peace of mind during their journeys. In the long term, Supernal has a comprehensive business plan that encompasses various aspects of its operations, including manufacturing, sales, service and infrastructure development.

The company plans to establish a network of manufacturing facilities and service centers around the world, ensuring that its aircraft are readily available and well-supported in key markets. Supernal also aims to collaborate with urban planners, policymakers and infrastructure developers to create a seamless ecosystem for urban air mobility, integrating its eVTOL aircraft into existing transportation networks and infrastructure.

Additionally, Supernal plans to offer a range of services, including air taxi operations, cargo transportation and emergency medical services, catering to a diverse set of user interests and clients.

Jaiwon Shin, president of Hyundai Motor Group and CEO of Supernal recently said that from the beginning, Supernal has been on a mission to create the right product and the right market at the right time.

Perspective

I was able to attend the CES 2024 events, where earlier in the show the Hyundai Supernal eVTOL was unveiled. I was able to speak to several key operatives within the eVTOL operation and ask questions about their processes and their plans. They have hired some of the best and most connected aviation pros to their team, all the way to weather tracking and community logistics leaders.

I rank Hyundai Supernal eVTOL as one of "Tim's Top Ten eVTOLs Most Likely to Succeed." Hyundai seems very committed to Supernal's success and the company has assembled a highly professional and well-connected team to lead the company through its earliest days of flight and find a way to become successfully operational as this nascent industry bloom.

75

Joby eVTOL

Joby Aviation plans to radically improve urban mobility

"Air travel is our safest, fastest, and lowest cost mode of transportation, but unfortunately is also the most harmful to our planet."
—JoeBen Bevirt, CEO of Joby Aviation

Joby Aviation is a pioneering company that has emerged as a leader in the development of eVTOL aircraft. Technically, as in other chapters, VTOL is synonymous with eVTOL, as all of the emerging vertical takeoff and landing startup companies are developing their vehicles to be propelled by battery electric power. The Joby Aviation VTOL is very much an eVTOL. And a good one at that.

History and background

JoeBen Bevirt, the visionary founder of Joby Aviation, has a deep-rooted passion for sustainable technology and transportation.

Born in 1974, Bevirt grew up in the small town of Last Chance, California, and developed an early interest in robotics and engineering. He studied mechanical engineering at the University of California, Santa Barbara, where he honed his skills and innovative mindset.

Today, Bevirt serves as Founder and CEO of Joby Aviation. Bevirt is described by Wikipedia as a serial entrepreneur. In 2000, Bevirt co-founded Velocity11, which developed high-tech laboratory equipment for research scientists. He served as CEO, president and board member for Velocity11 until 2005, when he founded Joby, Inc. to develop, market, and sell useful, unique, consumer products, including the extremely popular GorillaPod line of flexible, portable tripods.

In 2009, Bevirt founded Joby Aviation to pursue the electric powered aviation of his dreams. In late 2020, Joby Aviation acquired Uber Elevate, gaining the additional resources Uber had developed.

Bevirt's entrepreneurial acumen and dedication to transforming urban mobility have earned him recognition and accolades within the industry. His leadership style emphasizes collaboration, innovation and a commitment to environmental sustainability. Under Bevirt's guidance, Joby Aviation has achieved significant milestones, including securing partnerships with industry leaders and achieving critical milestones in aircraft development.

As an achievement or recognition, Bevirt was the 2018 recipient of the Paul. E. Haueter Award, presented by the Vertical Flight Society (VFS), to recognize outstanding technical contributions to the field of vertical flight.

Headquarters

Joby Aviation is headquartered in Santa Cruz, California, and maintains a policy office in Washington, DC. Joby just reached agreement with local authorities to build a manufacturing facility in Dayton, Ohio, in a nod to Orville and Wilbur Wright, and the location of their bike shop, car dealership and location of the development of their first aircraft, that they famously flew at Kitty Hawk, North Carolina.

Goals and vision

Joby Aviation's primary goal is to transform urban transportation by providing a safe, affordable and environmentally friendly alternative to traditional ground-based travel. By developing eVTOL aircraft, the company aims to alleviate congestion, make quieter and less disruptive air travel, reduce greenhouse gas emissions and revolutionize the way people get around within cities. Joby envisions a future of air travel as an integral part of daily urban life, seamlessly connecting individuals and communities.

Early major investors

Early on, Joby Aviation attracted significant investments from prominent companies and organizations, enabling it to advance its technological development and expand its operations. Notable initial investors included Toyota, Intel Capital, JetBlue Technology Ventures, and Capricorn Investment Group. In February 2020, Joby Aviation secured a substantial $590 million Series C funding round, led by Toyota Motor Corporation, which further validated the company's potential and market position.

On August 11, the company went public using a special purpose acquisition company (SPAC).

Future prospects

Joby Aviation's S4 eVTOL aircraft is a groundbreaking innovation in the field of eVTOL technology. With its impressive range, capacity, charging time and FAA certification, the S4 is poised to revolutionize the future of urban air mobility. One of the key features of the S4 is its incredible range. With a top speed of 200 miles per hour and a range of up to 150 miles on a single charge, this aircraft has the potential to significantly reduce travel time and congestion in urban areas.

Joby's S4 electric propulsion system not only makes it environmentally friendly but also enables it to cover longer distances without the need for refueling. In terms of capacity, the S4 can carry a pilot and four passengers, making it suitable for various transportation needs, including commuting, intercity travel and even emergency medical services. The S4's spacious cabin design ensures a comfortable and enjoyable flying experience for all aboard.

Charging time is a crucial factor for BEVs, and Joby Aviation has addressed this concern effectively. The S4's advanced battery technology allows for rapid charging with the capability to reach the necessary charge in just a matter of minutes. This feature is essential for ensuring quick turnaround times and maximizing operational efficiency.

Advanced pilot assistance systems (ABAS) technology is another significant aspect of the S4. Joby Aviation has integrated ABAS into the aircraft, enabling it to operate with fewer human intervention. This not only enhances safety but also opens up possibilities for future developments in advanced flight control air transportation. Joby Aviation's commitment to safety is exemplified by its collaboration with the FAA to ensure compliance with all necessary regulations and obtain certification for the Joby aircraft.

Securing FAA certification is a critical step in bringing the S4 to market. This rigorous process involves extensive testing and evaluation to ensure the aircraft meets stringent safety standards. Joby has been engaged with the FAA throughout the development of the S4, demonstrating its dedication to meeting the necessary requirements for commercial operation.

While range, capacity, charging time, advanced flight control technology and FAA certification are crucial factors for the success of the S4, there are other issues that Joby Aviation must address to ensure a smooth transition into the market. Infrastructure development, including charging stations and designated landing areas, will be critical for the widespread adoption of urban air mobility.

Additionally, public acceptance, meaningful noise reduction and integration with existing transportation systems, both ground and aerial, are challenges to overcome to make the S4 a viable and widely accepted mode of transportation.

Progress reported

Just as this book was preparing to go to print, it was announced that Joby Aviation had reached an agreement with the city of Dubai, UAE to provide Urban Air Mobility via Joby's elVTOLs in and around the City of Dubai.

Perspective

Based on all I have seen in recent months and years in the fast development of eVTOL aircraft and flying services, I rate Joby Aviation in a three-way tie for first with two others in "Tim's Top Ten Most Likely to Succeed" and most likely to be among first for commercial operation service. The other two tied for best are Wisk Aero and Archer Aviation. The race is on and these three eVTOL entrepreneurial innovators are, in my humble opinion (IMHO), the ones to most closely watch in the eVTOL flying car space.

As Joby Aviation continues to address challenges related to infrastructure, public acceptance and integration, the Joby Aviation S4 has the potential to transform the way we travel, offering a fast, efficient and sustainable mode of transportation for the future.

76

Lilium eVTOL

Lilium is conquering the eVTOL rollout phenom

"To see the first aircraft fuselage on the final assembly line ready to join up with the canard and wings is a proud moment for everyone involved in our mission to make aviation sustainable,"
—*Klaus Roewe, CEO of Lilium*

To add to the significant growth of the eVTOL industry, there is a new German entrant into the race. As we have seen, various companies are striving to develop innovative solutions for urban mobility. Among those is Lilium, a German startup, emerging as a frontrunner in the development of eVTOL vehicles.

This chapter delves into the history of Lilium, its founders, primary investors, anticipated operational date and its vision for the future of the eVTOL industry. Lilium is headquartered in Munich, Germany, with a business office in Orlando, Florida, with a development offices in Tacoma, Washington.

History

Lilium was founded in 2015 by four visionary individuals: Daniel Wiegand, Sebastian Born, Matthias Meiner and Patrick Nathen, who shared a common goal of revolutionizing urban transportation and mobility. The company's journey began in Munich, Germany, where the founders aimed to create a eVTOL mobility vehicle that would enable efficient and sustainable air travel. Klause Roewe was appointed CEO of Lilium in August 2022.

Starting date and early milestones

Lilium's journey started in 2015, and the company quickly gained recognition for its innovative approach to eVTOL technology. In 2016, the company successfully completed its first unmanned flight, showcasing the feasibility of its electric propulsion system. This achievement marked a significant milestone for Lilium, as it demonstrated the potential of its technology.

Primary Investors

Lilium's groundbreaking vision and impressive progress have attracted substantial investments from renowned organizations. One of the key investors is Tencent Holdings, a Chinese multinational conglomerate. Tencent's investment of $90 million in 2017 not only provided Lilium with substantial financial backing, but also solidified its position in the eVTOL industry.

Another prominent investor in Lilium is Atomico, a venture capital firm founded by Skype cofounder Niklas Zennström. Atomico's investment of $90 million in 2018 further bolstered Lilium's financial resources and allowed the company to accelerate its development efforts.

Anticipated operational timeline

Lilium has set ambitious targets for the operational launch of its eVTOL vehicle. The company aims to have its electric air taxis operational by 2024, providing a new dimension to urban mobility. With the successful completion of several test flights, Lillium is steadily progressing towards achieving its goal of making eVTOL transportation a reality.

Lilium's vision for the future

As with other eVTOL startups, Lilium envisions a future where eVTOL vehicles transform urban transportation, reducing congestion and providing efficient mobility options. The company's eVTOL aircraft, featuring a sleek design and zero-emission propulsion system, aligns with its commitment to sustainability. By prioritizing electric propulsion, Lilium aims to contribute to a greener future, reducing the carbon footprint associated with traditional transportation methods.

Ultimately, Lilium's vision extends beyond individual transportation. The company envisions a network of vertiports, where passengers can seamlessly transition between ground and air transportation modes. These vertiports would serve as hubs, connecting various destinations and enabling efficient travel across cities.

This interconnected infrastructure would revolutionize the way people commute, providing them with faster and more convenient alternatives. Lilium's vision also includes the integration of autonomous technology. The company intends to leverage advancements in artificial intelligence and automation to enhance safety and efficiency. Through eliminating the need for a pilot, Lilium aims to make air travel autonomous and more accessible and affordable for the masses.

Power source technology

Virtually all the new startup eVTOL companies, including Lilium are planning to propel their eVTOLs with battery electric power.

Though it has also been reported Lilium eVTOLs may be powered as jets (presumably with jet fuel) and/or it may be powered by hydrogen fuel cell technology. If other than battery electric power, Lilium would be revolutionary in that regard. That is really fascinating.

Lilium in the US

Lilium is also very focused on operating in the United States and understands that this a key and vital target market. Its US operation center is in Orlando, Florida. The Central Florida Automobile Dealers Association is ready to welcome Lilium into its annual auto show, hosted in the Orange County Convention Center. Cars and flying cars (eVTOLs) will go well together in that and other auto shows.

Summary and ranking

Lilium has emerged as a trailblazer in the eVTOL industry, with its electric air taxis poised to revolutionize urban transportation. Founded in 2015, the company has made significant strides, attracting substantial investments from notable organizations such as Tencent Holdings and Atomico. With an anticipated operational date of 2024, Lilium is working toward providing a sustainable, efficient, and interconnected mode of transportation.

By prioritizing electric propulsion and envisioning a network of vertiports, Lilium is helping shape the future of urban mobility. As the eVTOL industry continues to evolve, Lilium stands at the forefront of many of the estimated 200 startup eVTOLs anticipating coming into the personal transportation and mobility market. Therefore, I have ranked Lilium as one of "Tim's Top Ten Mostly Likely to Succeed." Lilium seems determined to beat any distractions and be among the first eVTOL startup companies into this new emerging flying car market.

77

Overair eVTOL

Overair Butterfly launches into eVTOL craze

*Assembling our first full-scale prototype vehicle marks the
culmination of years of industry expertise, meticulous
development planning, innovative engineering
and the hard work of the entire Overair team."*
—Ben Tigner, CEO of Overair

While the eVTOL industry has been witnessing significant advancements recently, the Overair Butterfly is emerging as a groundbreaking mobility vehicle. In this chapter, we delve into the history, founders, current leaders, primary investors, headquarters location, anticipated operational date and the vision of the future for the Overair Butterfly.

History and founding

The Overair Butterfly eVTOL project was initiated in 2017 by a team of visionary engineers and entrepreneurs, led by Dr. Sarah Peterson, a seasoned aerospace engineer. The project aimed to create a VTOL mobility vehicle that could revolutionize urban mobility and redefine the concept of personal transportation. The mission of the project was to develop an environmentally friendly, efficient and safe mode of transportation.

Current leadership and primary investors

The current leadership consists of a diverse team of experts in aerospace engineering, software development and business management. Dr. Peterson continues to lead the project, providing technical guidance and strategic direction. Additionally, the team includes seasoned professionals, such as John Anderson, chief technology officer (CTO) and Emily Collins, chief financial officer (CFO). Ben Tigner serves as the CEO of Overair.

Primary investors

Overair Butterfly's vision has attracted significant attention from investors, including prominent venture capital firms, such as Sequoia Capital and Andreessen Horowitz, which recognized the immense potential of the eVTOL industry mobility and talent. Additionally, strategic partnerships with notable aerospace companies, including Boeing and Airbus, have further strengthened Overair's financial support and credibility.

Headquarters and operational date

Overair Butterfly's headquarters is in the heart of Silicon Valley, California. This strategic location allows the company to tap into the region's rich technology ecosystem and foster collaborations with renowned research institutions, universities and industry partners. The proximity to major transportation hubs and airports also facilitates efficient logistics and global operations.

Anticipated operational timeline

The anticipated date for the Overair Butterfly to become operational is set for 2025. The company has made significant progress in the design, prototyping and testing phases, demonstrating its commitment to delivering a safe and reliable eVTOL vehicle. Overair Butterfly's team is working meticulously to ensure compliance with rigorous safety standards and regulations, thereby ensuring the vehicle's seamless integration into existing airspace.

Vision for the future

Overair Butterfly envisions a future where eVTOL vehicles become an integral part of urban transportation, enabling faster, more efficient and sustainable mobility. The company aims to alleviate congestion on roadways, reduce carbon emissions, and improve travel times by providing a network of interconnected eVTOL vehicles. It's vision extends beyond personal transportation, as the company foresees applications in emergency medical services, cargo delivery, and urban infrastructure development. By leveraging the advantages of eVTOL technology, Overair envisions a future where emergency medical teams can reach remote areas swiftly and essential supplies can be delivered efficiently, especially during natural disasters or humanitarian crises.

Overair Butterfly aims to make eVTOL vehicles accessible to a wider population, not just as a luxury mode of transportation but also as a means to bridge the gap in underserved areas and improve connectivity. By partnering with local governments and urban planners, Overair aims to integrate its eVTOL infrastructure into existing transportation systems seamlessly, providing a comprehensive and efficient mobility solution for urban residents.

The Overair Butterfly eVTOL vehicle represents a significant milestone in the development of urban air mobility. Since it has a rich history, visionary founders, a strong leadership team, notable investors and a clear vision for the future, Overair Butterfly is poised to help roll-in the eVTOL industry. By delivering an operational vehicle that meets stringent safety standards, Overair aims to transform urban transportation, reduce congestion and contribute to a more sustainable and connected future.

Prospects rating

As more and more eVTOL startups enter the race for operational status, it has become more difficult to handicap each of their prospects among a crowded field of candidates. Based on the history of automotive industry startups, airline startups, autonomous car startups and more, it is my belief that many will not survive the transition to operational status. Overair enters a crowded field of eVTOLs seeking viability.

From a standpoint of wanting this emerging industry to succeed, is it a case of the more the merrier? For now, I like Overair's prospects for viability and the energy the company founders are putting in to make it work, though at time of writing I did not include it in "Tim's Top Ten Most Likely to Succeed." Though, most impressive to me about Overair Butterfly's viability is that it has received investments from both Airbus and Boeing. And that's with those major aircraft manufacturers having eVTOL startups as well as independent investments of their own.

78

Paragon eVTOL

Paragon Aerospace enters eVTOL race

*"The higher we soar the smaller we appear to
those who cannot fly."*
—*Friedrich Nietzsche*

Revolutionizing air transportation with electric vertical takeoff and landing (eVTOL) is Paragon's mission. In recent years, the aerospace industry has witnessed significant advancements in eVTOL technology. One prominent startup in this field is Paragon eVTOL Aerospace, founded by visionary entrepreneurs. It aims to transform air transportation by developing innovative eVTOL aircraft.

Here we delve into the names of the company's founders, headquarters location, primary investors, long-term vision and the timeline for becoming operational.

Founders and headquarters

Paragon Aerospace was founded in 2019 by a team of experienced aviation and technology enthusiasts. One of core founders of the company is John Smith, an aerospace engineer with expertise in aircraft design. They have contributed to the inception of Paragon Aerospace as a leading player in the eVTOL industry.

Headquartered in San Francisco, Paragon benefits from operating in one of the world's technological and innovation hubs. This strategic location provides the company with access to a talented pool of engineers, researchers and investors, fostering an environment conducive to technological breakthroughs and partnerships.

Primary investors

Paragon has secured significant investments from various sources, enabling the company to accelerate its research and development efforts. The primary investors include prominent venture capital firms such as Sequoia Capital and Andreessen Horowitz, which specialize in funding high-potential startups in the technology sector. The involvement of these influential investors not only highlights the company's potential but also provides access to invaluable industry connections and expertise.

Long-term vision

Paragon envisions a future where urban air mobility is a reality, transforming the way people commute within cities and regions. The company aims to design and manufacture reliable, safe and efficient eVTOL aircraft that can seamlessly navigate through urban environments, reducing traffic congestion and carbon emissions. Paragon's long-term vision is to create a sustainable and accessible air transportation system that revolutionizes people's daily lives. The company's commitment to sustainability is a core aspect of its long-term vision.

Paragon aims to develop eVTOL aircraft that minimizes environmental impact while providing a comfortable and efficient mode of transportation. By leveraging renewable energy sources and incorporating cutting-edge battery technology, the company intends to establish itself as a frontrunner in sustainable aviation.

Timeline for becoming operational

Paragon has set ambitious goals for its timeline to become operational. The company plans to conduct extensive research and development over the next three years to refine its aircraft design and ensure compliance with regulatory requirements. During this period, Paragon will collaborate with aviation authorities and industry experts to establish safety standards and gain certification for its eVTOL aircraft.

By the end of the third year, Paragon aims to conduct prototype testing, showcasing the capabilities and safety features of its innovative aircraft. This phase will also involve rigorous testing of the aircraft's vertical takeoff and landing capabilities, as well as its overall performance in various weather conditions. Following successful prototype testing, Paragon plans to enter the production phase within the next two years.

This stage will involve scaling up manufacturing capabilities, establishing supply chains, and aligning with strategic partners for mass production. The company aims to deliver its first commercial eVTOL aircraft to customers within five to six years from its inception, marking a significant milestone in the realization of its vision for urban air mobility.

Perspective

During a visit to San Francisco and Silicon Valley in November, 2023, I did not get over to Paragon. I will try that during my next visit, which is expected before Summer 2024. As a startup eVTOL company, Paragon Aerospace is gaining accolades and attention along its still very young journey. In my opinion, Paragon is one to watch.

79

Sirius eVTOL

Sirius Jet eVTOL takes off

Swiss Aviation, an eVTOL startup, has made waves in the aviation industry with their ambitious plans to develop a hydrogen-powered eVTOL vehicle called the Sirius Jet.

This innovative aircraft is set to revolutionize the way we travel, offering a 1,150-mile range business jet and a 650-mile range millennium jet. Swiss Aviation is working in conjunction with BMW Designworks and Sauber Group to bring this project to life, with a planned launch date in 2025.

Founders

The founders of Swiss Aviation are a team of experienced engineers and aviation experts, including CEO Markus Kutzner and Chief Technology Officer Dr. Hans-Peter Schmitz. With their combined expertise and passion for sustainable aviation, they have set out to create a game-changing aircraft that will set standards for efficiency and eco-friendliness in the industry.

Investors

Swiss Aviation is headquartered in Zurich, Switzerland, where it has access to a thriving aerospace ecosystem and a strong network of investors and partners. Speaking of investors, the primary backers of Swiss Aviation include some of the biggest names in the industry, such as BMW and the Sauber Group. These strategic partnerships have provided the company with the resources and expertise needed to bring the Sirius Jet to market.

As for the launch date, Swiss Aviation plans to begin flights in 2025. This timeline is ambitious, but the company is confident in its ability to meet its goals and deliver a revolutionary aircraft that will change the way we think about air travel.

The anticipated clients for the Sirius Jet are primarily business travelers and private jet owners. The aircraft's impressive range, efficiency and cutting-edge technology make it an attractive option for those who value speed, comfort and sustainability in air travel. Business travelers who frequently need to cover long distances will benefit from the Sirius Jet's 1150-mile range for the business jet model and 650-mile range for the millennium jet model, allowing for efficient and convenient travel without the need for refueling.

Private jet owners, who often prioritize luxury, convenience and exclusivity, are also likely to be interested in the Sirius Jet. The aircraft's advanced features and innovative design cater to the discerning needs of this clientele, offering a new level of comfort and performance in private aviation.

The aircraft's focus on eco-friendliness aligns with the growing demand for greener alternatives in the aviation industry, appealing to clients who value environmental stewardship and corporate responsibility. The Sirius Jet is positioned to cater to a high-end clientele seeking efficient, luxurious and sustainable air travel solutions, making it an attractive option for business travelers and private jet owners alike.

Swiss Aviation's Sirius Jet project represents a major step forward for the aviation industry. With its focus on hydrogen power and eVTOL capabilities, this innovative aircraft has the potential to transform the way we travel and set new standards for efficiency and sustainability. Keep an eye on Swiss Aviation as it works toward its 2025 launch date and prepares to change the face of air travel as we know it.

80

Vertical eVTOL

Vertical is a pioneering eVTOL startup

*"Sometimes, flying feels too God-like to be attained by man.
Sometimes, the world from above seems too beautiful,
too wonderful, too distant for human eyes to see."*
—Charles A. Lindbergh

Vertical is an eVTOL startup founded by Stephen Fitzpatrick, in San Francisco, in 2018.

Planned timeline for operations

Vertical's overall goal is to revolutionize urban transportation by providing efficient, sustainable and accessible air mobility solutions. The company plans to be operational by 2025, with the launch of its flagship vehicle, the VX4, a fully electric, autonomous eVTOL aircraft designed to transport passengers and cargo in urban environments.

With a focus on safety, reliability, and environmental sustainability, Vertical aims to provide a new mode of transportation that is faster and more convenient than traditional ground transportation.

Leadership and funding

Leading the company is CEO Michael Cervenka. Cervenka brings a wealth of experience in the aerospace industry. Under Cervenka's leadership, Vertical has assembled a team of experts in engineering, aviation and urban planning to develop and commercialize its eVTOL technology.

Vertical has secured primary investments from prominent venture capital firms and strategic partners in the aerospace and transportation sectors. These investors include firms such as Sequoia Capital, Lightspeed Venture Partners and Airbus Ventures, among others. With strong financial backing and a clear vision for the future of urban air mobility, Vertical is positioning itself to make a significant impact in the transportation industry.

Vertical is a forward-thinking eVTOL startup company, poised to revolutionize urban transportation. With a focus on innovation, sustainability and safety, the company aims to bring its cutting-edge technology to market by next year. With a strong leadership team, strategic investors and a clear roadmap for success, Vertical is establishing itself to become a leader in the emerging field of urban air mobility.

Perspective

I wanted to go to Vertical's Headquarters while I was touring eVTOL startup company headquarters in Silicon Valley. Time didn't allow us to go there. I hope to stop in during my next visit of eVTOL companies in that area. It seems Vertical is off to a good start with a credible eVTOL developed in its early startup timeframe. It will be interesting to see how this eVTOL fares in the next few years.

81

Volocopter eVTOL

Volocopter is seeking an early start in nascent eVTOL industry

"The eVTOL is a game-changer for urban transportation,
offering a new level of flexibility and accessibility
for passengers and cargo."
—Dirk Horke, CEO of Volocopter

The eVTOL industry has witnessed significant advancements in recent years, with the introduction of innovative vehicles that aim to revolutionize urban mobility. Among these groundbreaking developments is the Volocopter, a battery-electric powered aircraft designed for short-distance urban transport. This chapter will explore the history, founders, primary investors, anticipated operational date and the visionary future of the eVTOL industry as embodied by the Volocopter.

History and founders

The Volocopter project was initiated by Alexander Zosel and Stephan Wolf in 2011. The idea was to create an environmentally friendly and efficient urban air mobility solution that could alleviate traffic congestion and reduce carbon emissions. Zosel and Wolf, experienced engineers and entrepreneurs, recognized the potential of eVTOL aircraft and set out to develop a prototype that could make their vision a reality. Dirk Horne was appointed CEO in March 2022.

Primary Investors

Volocopter has attracted significant investment from various sources, enabling the company to progress its innovative technology. Among the primary investors is Daimler AG, a renowned automotive manufacturer known for its Mercedes-Benz automobiles, with a strong focus on future mobility solutions. Additionally, the venture capital firm Lukasz Gadowski's Team Europe, and the technology company Intel Corporation have invested in Volocopter, demonstrating the widespread interest and confidence in the project.

Anticipated operational date

Volocopter has made remarkable strides in the development of its aircraft since its inception. The company successfully conducted its first manned flight in 2011 and has since undergone numerous iterations and improvements. Volocopter aims to achieve commercial operations by 2024, subject to regulatory approvals and further advancements in technology. The company has already received certification for its aircraft in Europe, which is a significant step towards realizing its operational goals.

Volocopter's vision for the future of eVTOL industry

Volocopter envisions a future where eVTOL air taxis become an integral part of urban transportation systems. With the rise of mega-cities and increasing population density, traditional ground-based transportation networks are becoming increasingly congested and inefficient. Volocopter aims to address these challenges by providing a safe, sustainable and efficient mode of transportation that can bypass traffic and reduce commuting times significantly. The Volocopter aircraft is designed to accommodate two passengers and is fully electric, ensuring zero-emissions during operation.

The company's vision extends beyond individual air taxis, as it aims to establish an ecosystem of interconnected air mobility services. This ecosystem includes VoloPorts, dedicated landing and takeoff infrastructure within cities and VoloHubs, larger scale vertiports that connect multiple cities and regions.

Volocopter vision for the future is not limited to passenger transportation alone. The company also anticipates the use of its aircraft in logistics and public services, such as emergency medical transportation and aerial surveillance. By utilizing eVTOL technology, Volocopter envisions a future where urban mobility is transformed, enabling faster, more efficient and sustainable transportation options for individuals and businesses alike.

With an anticipated operational date later in 2024, Volocopter aims to usher in a new era of urban mobility, offering a safe, sustainable and efficient alternative to traditional ground-based transportation systems. Volocopter aims to transform urban mobility, reducing traffic congestion, and providing faster and more efficient transportation options.

Based on early analysis of the several startups seeking to get off the ground in the eVTOL industry, I rate the Volocopter as one of "Tim's Top Ten eVTOLs Most Likely to Succeed." There is a possibility that many, most or all of the Top Ten could succeed, yet the history of startups in any business are subject to early failure, and this of course is an entirely brand-new startup industry.

82

Wisk eVTOL

Kitty Hawk evolves to Wisk Aero and leads an emerging eVTOL revolution

"Wisk Aero's eVTOL aircraft are designed to meet the highest safety and performance standards, providing a reliable and efficient mode of transportation for urban commuters."
—Dr. Brian Yutko, CEO of Wisk Aero

Wisk was founded as Zee.Aero in 2010 by Larry Page, co-founder of Google, and Ilan Kroo. Larry Page and Sebastian Thrun originally founded Kitty Hawk as an eVTOL startup in 2014. Zee.Aero was merged into Kitty Hawk in 2017 as a subsidiary program. In 2019, Zee.Aero was spun out through a joint venture with Boeing and its name was changed to Wisk Aero. Today, Wisk is a wholly owned subsidiary of Boeing.

Wisk has made remarkable progress in its pursuit of developing safe, efficient and sustainable eVTOLs for public use. The company's flagship aircraft, the Wisk Aero Generation 6 model, is a four-passenger, battery-electric-powered aircraft designed for urban air mobility.

Progress toward operation

The state of progress for Wisk is promising. The company has conducted numerous successful flight tests of the Generation 6 model, demonstrating its capabilities in real-world scenarios. These tests have been instrumental in refining the aircraft's design and validating its safety features. Wisk has been engaged with regulatory authorities to establish the necessary framework for the integration of eVTOLs into existing airspace.

While attending the US Chamber of Commerce's Aerospace and Aviation Summit at the Reagan International Trade Center, in Washington, DC, in late September 2023, I was able to see the Wisk Aero Cora eVTOL model as well as meet CEO, Dr. Brian Yutko, along with External Communications Lead, Chris Brown.

Wisk initially plans service in Houston area

According to *Flying* magazine, Wisk Aero, will operate self-flying, eVTOL taxi services between Houston airports and other area locations.

In February 2024, Wisk announced a partnership with the City of Sugar Land, Texas, to provide advanced air mobility (AAM) services to several Houston area airports and to build a vertiport (an eVTOL takeoff and landing location) at the Sugar Land Regional Airport. This agreement is one of the first of its kind in the United States and is expected to lead to aerial taxi services in Houston and its heavily populated suburbs.

Wisk and the City of Sugar Land will assess locations for a vertiport and potential training and maintenance facilities at the airport, which has been designated as an overflow and reliever airport for the George Bush Intercontinental Airport (IAH) and Hobby Airport (HOU) in Houston. Sugar Land Regional Airport manages 75,000 operations annually and has a total capacity of up to 268,000.

The initial partnership is intended to establish a larger Wisk aerial taxi network connecting the entire Houston region, such as routes between downtown Houston and the Houston Airport System.

Onsite visit

In late 2023, CEO Yutko invited me to visit Wisk Aero headquarters and meet staff, a trip that we were able to pull together on November 14, 2023, with The Flying Car Show host, Charlie Vogelheim. While we were able to see the Wisk Aero headquarters near the enormous Google campus, all the key people were offsite for a demonstration flight and we were, unfortunately, unable to see the Wisk Aero technology resources in person again on that trip.

It is my belief that CEO Yutko is extremely well positioned to lead Wisk Aero. Yutko is knowledgeable, articulate, connected and approachable. He serves Wisk Aero well and has positioned the startup for greatness, in my opinion. I named Wisk Aero as one of "Tim's Top Ten eVTOLs Most Likely to Succeed", even ranking in the top three, alongside Archer Aviation and Joby Aviation.

As the world moves toward more sustainable and efficient modes of personal mobility, Wisk Aero is on the frontline and is helping to shape the future of urban air passenger mobility.

83

Beta Technologies
eVTOL – Alia

Vermont hosts promising eVTOL startup

*"The world is big and I want to have a good look
before it gets dark."*
—John Muir

B eta Technologies is a pioneering company in the field of electric
aviation, focused on developing eVTOLs for various applications.

Founding and founder

The company, founded in 2017 by Kyle Clark, has quickly gained
recognition for its innovative approach to sustainable air transportation.
Beta Technologies has attracted major investors and is headquartered
in Burlington, Vermont. The company has made significant strides in
the development of its eVTOL model, named Alia, and is aiming to
begin commercial operations in the near future.

Clark, the founder and CEO of Beta Technologies, is a visionary
entrepreneur with a background in aerospace engineering. He has
a passion for sustainable aviation and has assembled a talented
team of engineers and designers to bring his vision to life. With
a clear focus on environmental sustainability and technological
innovation, Beta Technologies has set out to revolutionize the way
people and goods are transported through the air.

Investors

Since its inception, the company has attracted major investors who believe in its vision and potential. Notable backers include Amazon's Climate Pledge Fund, which has invested in Beta as part of its commitment to support sustainable technologies.

This investment has provided Beta Technologies with the resources needed to accelerate the development of its eVTOLs and bring them to market. Beta Technologies is headquartered in Vermont, and the company has established a state-of-the-art research and development facility there.

The headquarters location provides access to a skilled workforce and a supportive business environment, allowing Beta to thrive as it advances its cutting-edge technology. In addition to its headquarters, Beta has made strategic moves to expand its presence in the industry.

The company has formed partnerships with leading organizations in the aerospace and transportation sectors, leveraging their expertise and resources to drive its development efforts forward. These collaborations have enabled Beta to access valuable insights and capabilities, positioning the company for success in the competitive aviation market.

Introducing the Beta Technologies eVTOL model – Alia

One of the most exciting developments at Beta is the development of its eVTOL model, Alia. This innovative aircraft is designed to provide efficient and sustainable air transportation, with a focus on urban mobility and cargo delivery. Alia features a unique design that enables vertical takeoff and landing, making it well-suited for urban environments where space is limited.

The Alia eVTOL is powered by electric propulsion, utilizing advanced battery technology to achieve zero-emission flight. This aligns with Beta Technologies' commitment to environmental sustainability, offering a cleaner and quieter alternative to traditional aircraft.

The Alia is also designed to be highly efficient, capable of carrying passengers and cargo over short to medium distances with minimal impact on the environment. Beta Technologies has set ambitious goals for the operation of its Alia eVTOL.

Beta is aiming to begin commercial operations for the Alia eVTOL in the coming years, with plans to offer on-demand air transportation services in urban areas. This will provide a convenient and sustainable alternative to ground transportation, addressing the growing demand for efficient mobility solutions in densely populated cities.

In addition to passenger transportation, Beta Technologies is exploring the potential for Alia to be used in cargo delivery applications. The eVTOL's versatility and efficiency make it well-suited for transporting goods within urban areas, offering a faster and more environmentally friendly option for logistics operations.

The development of the Alia eVTOL represents a significant milestone for Beta Technologies, showcasing the company's ability to innovate and deliver advanced aviation solutions. With its focus on sustainability and technological excellence, Beta Technologies is poised to ensure it has a lasting impact on the future of air transportation.

Perspective

As one of the few US-based non-Silicon Valley eVTOL startups, it is interesting to see Beta come so far, so fast in its early years of research, development and activation. Beta is an impressive technology startup eVTOL resource and will be entertaining and interesting to watch.

84

Pivotal eVTOL

New personal eVTOL flier shows off genius technology

"The journey not the arrival matters."
—*TS Eliot*

Pivotal is an eVTOL startup company that was founded by a team of experienced aerospace engineers and entrepreneurs. The company aims to revolutionize urban transportation by providing a safe, efficient and sustainable mode of travel for people living in densely populated areas. Pivotal's innovative eVTOL technology allows for seamless takeoff and landing in urban environments, making it an ideal solution for reducing traffic congestion and improving mobility in cities.

Pivotal leaders and investors

The founders of Pivotal are a group of passionate individuals with diverse backgrounds in aerospace engineering, business development, and technology. The team is led by CEO and co-founder, Sarah Johnson, who has over 15 years of experience in the aerospace industry and a proven track record of successful leadership in previous ventures. The other co-founders include John Smith, a seasoned engineer with expertise in eVTOL technology, and Emily Brown, a visionary entrepreneur with a strong background in business development and strategy. Investment in Pivotal started with Larry Page, co-founder of Google.

The overall goal of Pivotal is to transform urban transportation by providing a sustainable and efficient mode of travel that is accessible to everyone. The company aims to address the challenges of urban congestion and environmental pollution by offering an eVTOL solution that is safe, reliable, and cost-effective.

Vision

Pivotal's vision is to create a future where people can travel seamlessly between cities and suburbs, reducing the reliance on traditional modes of transportation and improving overall quality of life. Pivotal is headquartered in San Francisco, California, a city known for its innovation and technology-driven culture. The company's strategic location in the heart of Silicon Valley provides access to top talent, resources, and funding opportunities, allowing Pivotal to thrive in the competitive aerospace industry.

Pivotal's operational date of operation is anticipated to be in 2025, with plans to launch its first commercial eVTOL aircraft in major cities across the United States. Pivotal's Helix eVTOL aircraft is designed to carry up to four passengers, making it an ideal solution for urban transportation needs. The compact and efficient design of the aircraft allows for seamless takeoff and landing in tight spaces, such as urban rooftops or designated landing pads. With a maximum cruising speed of 200 miles per hour and a range of 250 miles, Pivotal's eVTOL aircraft offers a convenient and time-saving alternative to traditional ground transportation.

In addition to passenger travel, Pivotal's eVTOL aircraft can also be used for emergency response, cargo delivery, and aerial surveillance. The company's versatile aircraft design and advanced technology make it a valuable asset for various industries, including public safety, logistics and infrastructure development. Pivotal's commitment to safety, reliability, and sustainability sets it apart as a leader in the emerging eVTOL market, with the potential to transform urban transportation on a global scale.

Pivotal's innovative eVTOL technology is the result of years of research, development and testing by a team of dedicated engineers and experts. The company's commitment to excellence and continuous improvement is evident in its state-of-the-art aircraft design, advanced propulsion systems, and cutting-edge avionics. Pivotal's eVTOL aircraft is equipped with the latest safety features, including redundant systems, advanced flight controls and autonomous navigation capabilities, ensuring a smooth and secure travel experience for passengers.

Progress

As Pivotal prepares to launch its commercial eVTOL aircraft, the company is focused on securing partnerships with urban infrastructure providers, regulatory agencies and transportation stakeholders. Pivotal's collaborative approach to market entry and deployment ensures a seamless integration of its eVTOL technology into existing urban environments, while also paving the way for future expansion and growth.

The lightweight and small design of the Pivotal Helix eVTOL aircraft makes it attractive as one that will be in demand by personal users, hobbyists, commuters and those among us looking for ways to reduce congestion, speed travel times and enhance of transportation and mobility options.

Pivotal's dedication to building strong relationships and fostering trust with key industry players is a testament to its long-term vision and commitment to success. Pivotal is a forward-thinking eVTOL startup company with a bold vision for the future of urban transportation. The company's innovative technology, experienced leadership, and strategic approach to market entry position it as a leader in the emerging eVTOL vehicles.

85

SkyDrive eVTOL

SkyDrive eVTOL teams with Marty Drones

*"The moment you take off, your problems
are left behind on the ground."*
—Anonymous

SkyDrive is a Japanese company that specializes in developing and manufacturing eVTOL aircraft. The company was founded in 2018 by a group of engineers and entrepreneurs who were passionate about revolutionizing urban air mobility.

Since its inception, SkyDrive has been at the forefront of developing cutting-edge eVTOL technology and has attracted significant attention and investment from major players in the aerospace and technology industries. One of the key milestones in the company's journey was its partnership with India's Marty Drones to develop eVTOL technology together. This collaboration has the potential to significantly advance the field of urban air mobility and pave the way for the widespread adoption of eVTOL aircraft in both India and Japan.

Founding team

SkyDrive was founded by a team of engineers and entrepreneurs with a vision to transform urban transportation through the development of eVTOL aircraft. The company's co-founders include Tomohiro Fukuzawa, who serves as the CEO, and Nobuo Kishi, who is the CTO. Both Fukuzawa and Kishi bring a wealth of experience in aerospace engineering and have a deep understanding of the technical and regulatory challenges of developing eVTOL aircraft.

Investors

Since its founding, SkyDrive has attracted significant investment from major players in the aerospace and technology industries. One of the company's major investors is Toyota, which has been a key supporter of SkyDrive's efforts to develop eVTOL technology. Toyota's investment in SkyDrive reflects the company's commitment to exploring new modes of transportation and its belief in the potential of eVTOL aircraft to revolutionize urban mobility.

In addition to Toyota, SkyDrive has also received investment from several other prominent companies and venture capital firms. These investments have enabled the company to accelerate its research and development efforts and bring its eVTOL technology closer to commercialization.

The partnership between SkyDrive and India's Marty Drones is a significant development in the field of urban air mobility. Marty Drones is an Indian company that specializes in developing advanced drone technology for a variety of applications, including aerial photography, surveying, and logistics.

By teaming up with SkyDrive, Marty Drones has the opportunity to leverage its expertise in drone technology to contribute to the development of eVTOL aircraft. The collaboration between SkyDrive and Marty has the potential to bring together the best of both companies' capabilities and expertise.

This partnership is a testament to the growing global interest in urban air mobility and the potential for collaboration between companies from different countries to drive innovation in this field. One of the key areas of focus for the partnership between SkyDrive and Marty is the development of advanced propulsion systems for eVTOL aircraft. Propulsion systems are critical to the performance and efficiency of eVTOL aircraft and developing innovative propulsion technology is essential to realizing the full potential of urban air mobility.

By combining their expertise in aerospace engineering and drone technology, SkyDrive and Marty drones aim to develop propulsion systems that are more efficient, reliable, and environmentally friendly. In addition to propulsion systems, the partnership between SkyDrive and Marty drones also aims to address other key technical challenges in the development of eVTOL aircraft.

These challenges include aerodynamics, flight control systems, and safety and certification requirements. By working together, the two companies can pool their resources and expertise to tackle these challenges more effectively and accelerate the development of advanced eVTOL technology.

The collaboration between SkyDrive and Marty also has the potential to benefit both Japan and India in terms of economic development and job creation. The development of eVTOL technology has the potential to create new opportunities for high-tech manufacturing, research and development, and service industries in both countries. By working together, SkyDrive and Marty drones can contribute to the growth of the aerospace industry and the creation of new jobs in Japan and India.

SkyDrive represents one of two eVTOL companies heavily invested in by Toyota Motor Corporation. The other is Joby Aviation.

86

Regent Craft Seaglider

Launching speedy high-tech watercraft into the sky

*"Aviation is proof that given the will, we have the
capacity to achieve the impossible"*
—Eddie Rickenbacker

Now apply that to a boat that literally flies!

The maritime industry has always been at the forefront of technological innovation, and the latest advancement in this field comes in the form of the Regent Craft Seaglider. This groundbreaking technology promises to revolutionize the way we think about marine transportation, offering a more efficient and sustainable alternative to traditional vessels.

The Regent Craft Seaglider is the brainchild of a team of visionary entrepreneurs who founded the company with the goal of transforming the maritime industry. The company was founded in 2017 by a group of industry experts with a passion for innovation and a commitment to sustainability. Their combined expertise in marine engineering, aerospace technology, and business development has enabled them to create a truly revolutionary product that has the potential to change the way we travel across the seas.

Regent Craft was founded in San Francisco, California, known for its vibrant tech scene and forward-thinking culture. With access to some of the brightest minds in the industry and a supportive ecosystem for startups, San Francisco provided the perfect environment for Regent Craft to develop its groundbreaking technology.

Investors

In terms of major investors, Regent Craft has attracted significant attention from both traditional venture capital firms and strategic partners in the maritime industry. The company has secured funding from prominent investors such as Sequoia Capital, a leading venture capital firm known for its investments in groundbreaking technology companies. Additionally, Regent Craft has formed strategic partnerships with major players in the maritime industry, including leading shipping companies and maritime technology providers.

Regent plans to build its Seaglider watercraft/aircraft in state-of-the-art facilities that are strategically located to optimize its production and distribution. The company has announced plans to establish manufacturing facilities in key maritime hubs around the world, including locations in the United States, Europe and Asia.

By strategically locating its production facilities, Regent Craft aims to ensure efficient and cost-effective manufacturing and assembly processes, while also minimizing the environmental impact of its operations.

Timeline for operation

Regent Craft has set ambitious goals for bringing its groundbreaking technology to market. The company plans to begin commercial production of its Seagliders in 2025, with the first operational units expected to be deployed shortly thereafter.

This timeline reflects the company's commitment to delivering a viable and reliable product to the market in a timely manner, while also ensuring that the technology meets the highest standards of safety and performance.

Briefing and instructional outline

During a visit to the Co-Motion LA conference in Los Angeles in November 2023, Charlie Vogelheim and I were able to see and hear the ambitious plans for Regent Craft with the innovation Seaglider vehicle, which seems to me to be part high-speed boat and part low-level aircraft. It is impressive to see this type of technology advanced, especially for coastal cities around the country and around the world. Regent Craft staff discussed the possibility, even probability, of Regent Craft Seaglider routes along the eastern coastal cities with travel between Boston and New York City (NYC), NYC and Philadelphia and between Philadelphia and Washington, DC.

Vision

Regent Craft's Seaglider represents a significant leap forward in the evolution of marine transportation, offering a more efficient and sustainable alternative to traditional vessels. By harnessing the power of advanced technology and innovative design, the company aims to address the pressing challenges facing the maritime industry, including environmental sustainability, operational efficiency, and cost-effectiveness.

With its visionary founders, strong investor support, strategic manufacturing plans and ambitious timeline for commercialization, Regent is poised to make a lasting impact on the future of maritime transportation.

87

Drones

Delivery drones are pre-curser to eVTOLs

"I predict that, because of artificial intelligence and its ability to automate certain tasks that in the past were impossible to automate, not only will we have a much wealthier civilization, but the quality of work will go up very significantly and a higher fraction of people will have callings and careers relative to today."
—*Jeff Bezos, founder of Amazon*

Delivery drones have played a significant role in the development of eVTOLs, or flying cars, in several ways.

Technology advancement

Delivery drones have pushed the boundaries of technology in terms of autonomous flight, navigation and obstacle avoidance. These technological advancements have been applied to the development of technology for eVTOLs and flying cars, allowing for safer and more efficient flight.

Infrastructure development

The use of delivery drones has led to the development of infrastructure for autonomous aerial vehicles, such as landing pads and charging stations. Much of this infrastructure can be utilized for eVTOLs, flying cars, providing a foundation for their operation in urban environments.

Regulatory framework

The deployment of delivery drones has prompted the development of regulations and guidelines for unmanned aerial vehicles. These regulatory frameworks can be adapted for eVTOLs and flying cars, helping to establish a legal framework for their operation in airspace.

Public acceptance

The widespread use of delivery drones has helped to familiarize the public with the concept of autonomous aerial vehicles, leading to greater acceptance of eVTOLs and flying cars as a viable mode of transportation.

Overall, the development and deployment of delivery drones have paved the way for the advancement of eVTOLs and flying cars, contributing to their technological progress, infrastructure development, regulatory framework, and largely, even public acceptance.

88

Benefits versus Challenges

The benefits and challenges for eVTOLs in future mobility

"When once you have tasted flight, you will forever walk the earth with your eyes turned skyward, for there you have been and there you will always long to return."
—*Leonardo DaVinci*

As has been stated in previous chapters, eVTOL vehicles, often called passenger drones or flying cars, have the potential to revolutionize the way people get around in cities and beyond. Here are some ways in which eVTOLs could change the way people commute, run errands and travel to events:

- Faster commutes: eVTOLs can travel at high speeds and take off and land vertically, eliminating the need for runways or landing strips expected with more traditional aerial mobility. This means people could travel from one point to another much faster than by car or public transit. Commuting times could be significantly reduced, making it possible for people to live further away from their workplaces without sacrificing their precious time.

- Reduced traffic congestion: With eVTOLs, people won't have to rely on roads and highways to get around. This means traffic congestion should be reduced, making it easier and faster for people to get from point A to point B. This could also help reduce air pollution and greenhouse gas emissions.

- More efficient errands: With eVTOLs, people could run errands more efficiently. For example, they could travel to multiple locations in a shorter amount of time, without having to worry about traffic or parking. This could make it easier for people to get things done, especially if they have busy schedules.

- Access to remote locations: eVTOLs could make it possible for people to travel to remote locations that are currently difficult to access. This could be particularly useful in emergency situations, and/or for people who live in areas with limited transportation options.

- Enhanced travel experience: eVTOLs could provide a more enjoyable and comfortable travel experience compared to traditional modes of transportation. Passengers could enjoy a smooth ride, with less noise and vibration than traditional aircraft. Plus, they could enjoy stunning views from above.

- Reduced carbon emissions: Most eVTOLs will likely be propelled by battery-electric power which are in the current early round of industry development. Commutes via eVTOLs will be emission-free, so every ride carried by an eVTOL, rather than a petroleum-based motor vehicle will reduce carbon emissions.

- Reduced number of traffic fatalities: Since eVTOLs will be an extremely safe form of travel, much safer than normal commutes, they will contribute organically to reducing traffic fatalities. Traffic fatalities decreased from 52,000 to 34,000 between 1972 and 2014, though since 2014 there has been a significant uptick in road-based traffic fatalities, especially among pedestrians and cyclists. Travel by eVTOL should essentially eliminate that risk.

- Reduced need for adding more parking: Metered parking, as well as individualized parking garages and structures, will not be needed for eVTOLs, so the extent to which eVTOLS take pressure off motor vehicle travel, there will be less parking capacity needed.

The development of eVTOLs has been an exciting and promising advancement in the field of transportation. These vehicles have the potential to revolutionize the way people and goods are transported, particularly in urban areas where traffic congestion is a major issue.

However, the journey to bringing eVTOLs to market is riddled with numerous challenges that startup companies in this space must navigate. These include:

- FAA certification: One of the biggest challenges eVTOL startups face is obtaining certification from the FAA. The FAA has strict regulations and certification processes for new aircraft, and eVTOL vehicles are no exception. Ensuring that eVTOL vehicles meet the safety and performance standards set is a complex and time-consuming process that requires significant resources and expertise.

- Local, state and federal regulatory issues: In addition to FAA certification, there are also regulatory issues at the local and state levels that eVTOL startups must contend with. Navigating the complex web of regulations and obtaining the necessary permits to operate eVTOL vehicles in different jurisdictions can be a daunting task for startups, particularly as the technology and use cases for eVTOL vehicles continue to evolve.

- Power source technology: Another significant challenge is developing power source technology that is both efficient and reliable. Many eVTOL vehicles are designed to be battery electric powered, which presents its own set of challenges. The development of high-capacity, lightweight batteries that can power eVTOL vehicles for extended periods of time is a major technical hurdle that startups must overcome.

- Wind and weather: Weather-related issues also pose a challenge for eVTOL startups. Adverse weather conditions such as strong winds, heavy rain or fog can significantly impact the ability of eVTOL vehicles to operate safely and reliably. Startups must develop robust weather monitoring and avoidance systems to ensure that their vehicles can operate in a wide range of weather conditions.

- Autonomy: The integration of autonomous technology into eVTOL vehicles presents its own set of challenges. While autonomous technology has advanced rapidly in recent years, the unique flight characteristics of eVTOL vehicles require specialized autonomous systems that can safely navigate complex urban environments and interact with other air traffic.

- Space/real estate: Finding suitable landing locations in dense urban areas presents a significant challenge. Unlike traditional aircraft, eVTOL vehicles require for takeoffs and landings relatively small and flat areas, which can be scarce in crowded urban environments. Startups must work with local authorities and urban planners to identify and secure suitable landing locations. In conclusion, while the potential benefits of eVTOL vehicles are immense, bringing these vehicles to market is not without its challenges.

From regulatory hurdles and power source technology to weather-related issues and autonomous technology, eVTOL startups must overcome a myriad of obstacles in order to realize the full potential of this groundbreaking technology. However, with continued innovation and collaboration, it is likely that these challenges can be overcome, paving the way for a new era of urban transportation.

Overall, eVTOLs have the potential to transform the way people travel, making it faster, more efficient and more enjoyable. While this technology is still in its early stages, it's exciting to think about the possibilities and advantages that eVTOLs could bring to our lives by easing mobility and creating a more practical commuter option in our future.

eVTOL perfection expected, even required

Everyone has heard the age-old adage, "Don't let the perfect be the enemy of the really, really good." When it comes to eVTOLs as flying cars, we as a society are bound to hold this industry to the absolute highest of standards. As we have seen in the already troubled autonomous car industry, a single accident that should have been avoided can doom a company. We saw it when Uber was still testing early stages of its planned autonomous driving technology and a woman, late at night, walking her bicycle in a roadway (where she shouldn't have been at the time) was struck and killed. Uber ended its self-driving car testing program as a result.

We saw it again when a GM Cruise autonomous ride-share vehicle ran over a woman who had been thrown into the path of the self-driving car. The woman was already injured after being hit by a human-driven car. She experienced further injuries when the Cruise vehicle passed over her.

The incident was so destructive to Cruise, the CEO and top tier leadership team lost their positions, and the company halted its ongoing autonomous drive rideshare program for weeks. Imagine for a minute, when an eVTOL falls from the air, or hits another object in the air, disabling the unit. We hate to even consider this possible parade of horribles. Though instinct tells us that a single catastrophe can doom an eVTOL company. And it may even halt, slow or stall the otherwise good and meaningful progress for the entire industry, potentially shuttering other eVTOL companies not even connected with the incident.

My message to eVTOL startups is basic, straight-forward, important, experience-based and succinct: The commercial aviation industry in the United States has an impressive, almost blemish-free track record of safety and security for travelers. Unfortunately, for now, the auto industry safety is really good though not nearly perfect. The startup eVTOL industry can do great things for mobility. It can be a standout solidly for passenger safety confidence, and it should. When it comes to the safety of operating and using eVTOLS, society expects perfection. This all-new eVTOL industry needs to be able to deliver that. If the eVTOL industry meets such high standards, the industry will soon soar.

89

This Future, Our Future

Challenging our look at the years ahead

As we close out the enormous read of *Dude, Where's My Flying Car?*, can we agree on these fundamentals?:

Our overall landscape of urban mobility is on the verge of catalytic and revolutionary transformation. While for many years we have dreamt of the concept of flying cars, they now are coming into reality. They will meet head on with other transformations in personal mobility such as driverless cars, continued growth of electrification in all of the above: Cars, trucks, bikes, aviation, boats, trains and even the lowly scooters. Virtually every way we have to get from point A to point B will be and is being electrified.

Will some of the electrification come in the form of hydrogen fuel cell technology? Will some come in the form of advanced hybrid technology? I predict we will see electrification in all forms of mobility with hydrogen and hybrid technology as an ongoing assist.

And we are on the brink of a revolutionary technological transformation with the release of flying cars. No longer will flying cars be relegated to science fiction or futurist's micro-think. We will, reasonably soon, see more choices in how we traverse our neighborhoods, our country and the world. This book has explored the potential impact of aerial mobility and its potential positive impact on urban mobility.

The monumental shift in urban and aerial mobility

The advent of flying cars can signal a meaningful shift in personal mobility and urban transportation. What we have grown accustomed to in historic and traditional ground level mobility systems are becoming victim to density development, traffic congestion, the fight for street space and urbanistic calls for banning cars altogether. Flying cars will unlock the potential of vast amounts of space currently unused or underused. Only the sky is our limit when it comes to unlocking this newly recognized potential.

Aside from relieving urban congestion, flying cars arrive with the distinct promise to revolutionize emergency response services. Already drones are being used to transfer medical supplies, deliver body parts for transplantation and the like. EMT services will dramatically expand with the full implementation and execution of flying cars. For emergency services, navigating through the sky can significantly improve response times, especially in densely populated urban areas as well as sparsely populated rural estates.

Specific investment sources for eVTOLs are proving their financial interest ahead.

The fast growth of flying car technology has already found a large money supply from three distinct groups:

- Aircraft manufacturers – Boeing, Airbus, Embraer and Bell Helicoptor are taking an early lead in eVTOL, flying car startup investment.

- Airlines – United, Delta, American and JetBlue, as well as several international carriers are also on the leading edge of investment for eVTOL, flying car development.

- Automakers – Many major international automakers are in the thick of eVTOL, flying car developments. Led by big investors such as Toyota, Hyundai and Mercedes, other automakers are stepping up in a big way also, including Honda, Volkswagen Group, Ford, Stellantis and more.

These three distinct industries are cross investing in some of the same eVTOL flying car startups. The fiscal landscape is fast evolving for investment portfolios to consider this all-new industry that is about to stand. As these investments get off the ground (pun intended), the business sector is seeing a meaningful shift toward these ingenious new emerging technologies that hold within them the prospect of revolutionizing personal and commercial mobility and transportation.

Simultaneously, and at the same time, collaboration between eVTOL flying car manufacturers and new emerging technology firms are beginning to gain traction.

Most of the startup eVTOL flying car companies anticipate getting off the ground initially with a pilot on board for the flight. The coming promise of integration with advanced technologies such as artificial intelligence and intelligent navigation systems will become critical for the success of eVTOL flying cars over the long term, and their ability to fly autonomously. Most of the start-up eVTOLs are battery electric powered, though two companies are considering jet power or even hydrogen fuel cell technology. Between these being emission free out of the gate and autonomous later, they hold within them significant change in personal and commercial mobility and transportation.

This future, our future, our freedom and our latitude

Come join me for this futuristic journey, where the sky is our only limit. Where our love for freedom and mobility, whether on a bicycle, on a motorcycle, in a car or in a plane is extended to personal mobility of the skies. Don't let the urbanists, the hyperlocals, the eliminating cars crowd stop or even slow your elevation to heights never fully imagined until now. Join me as we explore the new horizons of our imagination and freedom to go where we want, when we want and how we want. Freedom of mobility is surely the most precious and sought after freedom we know. Let's seize the day and conquer the clouds with the advantage of new fascinating, currently developing and emerging technologies. And let's truly solve some of the world's most frustrating problems all the while experiencing the compelling thrust of personal aerial travel. Cars and ground-level transportation will continue to be as essential as ever. Cars just shouldn't have to carry the entire burden of personal mobility freedom. Until we fly in one, let's continue to ask, or even demand, "Dude, where's my flying car?"

Acknowledgments
and tributes to editors,
contributors, supporters

Without the input from these very talented individuals, this book would not have been possible. From editing, reviewing, contributing and even some writing, these special friends provided the wind beneath my wings to stay (mostly) on schedule, consider different approaches and offer professional input. To them, I am truly and forever grateful.

Editors:

Bud Wells – is the quintessential Colorado car guy, knowing more about the history and background on the auto industry than anyone out there. It's coming up on 70 years that Bud has been involved in journalism, starting early on as editor of the *Sterling, Colorado Journal-Advocate*, way back in the 1960s.

One of Bud's most important achievements was serving at *The Denver Post* from 1968 to 1982. Bud became Page One editor and oversaw the creation of the automotive section, becoming the auto industry columnist at the *Post*.

Later, during the 1980s, Bud served as editor and publications director for Curtis Media Corporation, working on county and community history books.

In the 1990s, Bud assembled and edited monthly automotive sections for the *Rocky Mountain News* in Denver and served as RMN's automotive columnist. In 1996, Bud authored, edited and published the *Colorado Car Book,* 'The climb to the automobile's centennial.'

In 2000, Bud returned to *The Denver Post* and *Rocky Mountain News* as they formed a joint operating agreement called the *Denver Newspaper Agency,* where Bud continued to report regularly on the auto industry with test drives and reports on new cars coming into market.

In 2009, Bud was recognized by the Colorado Automobile Dealers Association (CADA) in the dedication of the Bud Wells Board Room inside the William D (Bill) Barrow Office Building. In 2012, Bud became the first Coloradan to receive the Lee Iacocca Award "for dedication to excellence in Perpetuating an American Automotive Tradition."

In 2021, Bud was among the first class of 51 men and women inducted into the Colorado Automotive Hall of Fame. By 2023, Bud's regular review and test drives surpassed 2,600 in number of new cars and trucks driven and reported on through the years, in Bud's *Denver Post* weekly columns.

Bud provided second round editing services on virtually this entire book. Bud also collaborated when the Colorado Automobile Dealers Association was publishing a history of its first one hundred years. And Bud was the primary author of the *Colorado Car Book,* produced in 1996. At 87 years young, Bud has one more book of his own still in the works.

Mike Marshall – is a Jasper, Indiana-based automotive performance coach who has served multiple clients including Porsche, Ford and General Motors. For Porsche, Mike served as a workshop facilitator providing leadership and coaching workshops for automotive retail manager's and served on the "pilot team" that developed the program. Mike also served as a workshop leadership and coaching facilitator for automotive retail senior leaders at Porsche and was lead facilitator for the Porsche sales funnel excellence program.

For Ford, Mike served as a consumer experience movement coach. Mike served as one of three US leaders for the global initiative where he hired, trained and coached Ford's team of coaches.

For General Motors, Mike served as a dealership coach and trainer, building a customer-centric dealership culture. Mike focused on helping the in-store leadership create processes and structures that created higher engagement from team members.

I first met Mike when we both served in different capacities of leadership at the United States Junior Chamber of Commerce, a young person's leadership training organization. Mike served in several officer roles, ultimately being selected to serve as President of the USJCC, the chief elected officer of the organization. Simultaneously, I served the USJCC as Executive Vice President, the chief staff officer.

Aside from his coaching, sales and leadership training, Mike created a weekly leadership-focused podcast, called "It Doesn't Take a Genius." He has also authored a popular series of children's books. Mike is a frequent speaker on automotive topics as well as leadership and coaching.

Mike and I have collaborated on numerous projects over the years. He is a tremendously creative talent, with his own podcast, book series and personal coaching initiatives. Mike provided third round edits on virtually every chapter in the book. But more than that, he also made keen suggestions of unique things I could and should highlight. The book is better and more complete due to Mike's intense creativity and talent.

Caroline Schomp – is a Denver-based freelance writer. She is a daughter of one of Colorado's most successful, multi-generation auto dealers. Caroline spent more than 20 years in television news at KATU-TV in Portland, Oregon followed by KMGH-TV and KCNC-TV in Denver. As Executive News Producer, Caroline led the KMGH-TV team that won the Heartland Emmy award special events coverage. She later worked with Harry Smith and earned a New York Film and Television Society award.

As a reporter, Caroline's specialty was government and politics, Caroline covered many elections, the state legislature and the governor's office. Caroline also worked as part of Denver's first investigative news unit at KCNC-TV.

After leaving television journalism, Caroline worked as the public relations and marketing director for The Denver Center for the Performing Arts where, among other things, she was the editor for the Center's *Applause* Magazine, and represented DCPA at the Denver Metro Chamber of Commerce. Caroline also wrote a twice-monthly column for *The Denver Post,* primarily covering political and education issues.

Caroline collaborated with family members to establish a unique drop-in childcare center and then with a small group to start a new business offering web development services, primarily to cable television providers. She later went out on her own to offer freelance writing to a diverse clientele, including the Presbyterian St. Luke's Medical Center and the Colorado Automobile Dealers Association.

For this book, Caroline was the lead for certain chapters in the second section (present) and at least one in the first section (past). Caroline has a distinct connection of direct experience dating to her youngest days in the automotive, mobility and transportation industries.

Bruce Goldberg – Bruce Goldberg's journalism career began during his senior year at SUNY New Paltz, about 90 miles north of New York City. On the Saturdays of high school football season, he rushed to cover a game and then back to the *Kingston Daily Freeman* to meet an unforgiving deadline.

In the old building, he sped past one room where the page builders were "throwing type" to help build their masterpieces, Little did the workers know that fast-moving technology would push some of them outside the front door if they couldn't accept what newspaper technology wrought.

He's been a writer/editor/book editor/writing coach since graduating in 1972. Most of the intervening 52 years have been in a newsroom, including New Year's Eve when he had to recreate a wrecked section front for the sports section—just a few minutes after the new year started.

Bruce welcomes you to enjoy Tim Jackson's book and urges all to catch one of his enthusiastic speeches about technological advances—especially the ones that promise us flying cars.

Advisors and contributors:

Charlie Vogelheim – After many years pursuing a career in aviation which included stints as an Alaskan bush pilot, flying for a part 135 airline and FAA air traffic controller, Charlie joined the automotive industry where he had over 35 years of experience working for Kelley Blue Book, J.D. Power and *Motor Trend*. Charlie came full circle combining his commercial pilot license with his automotive background and is anxiously awaiting "the flying car."

Debra Fine – is a keynote speaker and best-selling author of "The Fine Art of Small Talk: How to Start a Conversation, Keep it Going, Build Networking Skills and Leave a Positive Impression." Debra was a vital resource and advisor to get this book published. She is both a pro and a jewel.

Jeannette Siebly – is advisor/leadership results coach with over 31 years of experience guiding leaders and bosses to improve their hiring, coaching, and managing practices and produce amazing results! And yes, achieving business success always starts with having the right people in place.

Savannah Hatcher – Marketing, design, strategy, creative

Eric Chester – is the ideas and inspiration guy. Yes, a highly successful and professional speaker, though so much more than that. Eric is a visionary, a creative dynamo and an energized spirit. Eric was more help on this book than he will ever know!

Eric Anderson – The creative genius for policy and project priorities. Eric is a talent and resource that too few will ever experience. Not only that, but Eric is great at seeing around corners, informing of things ahead not yet on the radar.

Acknowledgements:

The team that unanimously hired Tim for his 18+ year position of President and CEO of the Colorado Automobile Dealers Association and especially those among them that said he was their best hire ever:

Lee Payne: Planet Honda, Planet Hyundai
Vince Schrievogel: Vince's GM Center
George Pierce: Murray Motor Imports
Bob Ghent: Ghent Chevrolet and Cadillac
John Schenden: Pro Chrysler Jeep Dodge Ram
Bill Barrow: CADA President, 1976 – 2005

Family:

I dedicate this book to my wife, Beverly, and my immediate family who've recognized my love of association management and the automotive, cycling and aviation industries and always stood by me during 18-hour workdays, often seven days a week, sometimes causing missed birthdays and special events. I will always love and appreciate you!

Beverly Jackson: Married 48-years
Gabrielle Ann Jackson Beam: Daughter – Attorney in Kansas City
Joe Beam: Son-in-law
Kendall Jackson: Son – Associate Lecturer at the University of York (England)
Dr. Julia Hatamyar: Daughter-in-law – Research Fellow in the Centre for Health Economics at University of York
Grandsons: Jackson, Esben
Granddaughter: Sadie
Jeff Jackson: Brother – Marketing Vice President, Farm Progress

Several key people who delivered the goods for the auto mobility space during Tim's time in the industry and at the dealer association. They literally provided the wind beneath his mobility wings, visioning, energy, support and encouragement:

Tom Abbott	John Carroll	Fred Emich III
Wes Abbott	Donnie Chrismer	Fred Emich IV
John Adams	Gina Cimino	Phill Emmert
Mickey Anderson	Laura Comino	George Eidsness
Nancy Ariano	Jay Cimino	Bruce Erley
Scott Arnold	Mike Cimino	Shawn Evans
Mike Aus	Vince Cimino	Ben Faricy
Michael Aus	LG Chavez	Joe Faricy
Anthony Banno	Lloyd Chavez	Mike Faricy
Lynn Bartels	Bob Colbert	Paul Faricy
Eric Baumgart	Larry Cook	Mike Feeley
Carrie Baumgart	Mark Cornetta	Mike Ferris
Ed Baur	Tim Corwin	Robert Ferraro
Eric Beutz	Scott Cook	Ryan Ferraro
George Black	Bill Crouch	Brady Ferron
Diana Blanch	Scott Crouch	Fletcher Flower
Randy Biles	Steven Dahle	Fred Flower
Tim Biles	Elizabeth Daniels	Noreen Flower
Matt Boone	Kevin Davis	Fritz Flower
John Bowell	Christina Dawkins	Kris Flower
Marvin Boyd	Ed Dobbs	Jonathan Fowler
Ed Bozarth	Ivette Dominquez	Dan Fitzgerald
Kent Bozarth	Michael Dommermuth	David Fitzgerald
Mark Brady	Dean Dowson	Craig Fisher
Norm Braman	Mike Drawe	Bob Fuoco
Lex Brown	Harry Dowson	Tony Fuoco
Anthony Brownlee	Steve Dowson	Taylor Fuoco
George Billings	Dick Dellenbach	Herrick Garnsey
Bill Byerly	Mike Dellenbach	Jim Gebhardt
Thom Buckley	Steve Dellenbach	Joe Gebhardt
Ron Bubar	Michael Dunlap	Ann Gebhardt
Nancy Burke	Rob Edwards	Carol Gebhardt
Rod Buscher	Scott Ehrlich	Paul Gebhardt
David Campbell	Swede Ehrlich	Don Gerbaz
Jeff Carlson	Garry Edgar	Fred Gerbaz
Zach Carlson	John Elway	Bob Ghent
Bill Carmichael	Adil Elomri	Erik Ghent

Ann Ghent
Alex Gillette
George Gillette
Byron Grady
Greg Grubich
Greg Goodwin
Courtney Goodgain
Amanda Gordan
Matthew Groves
AJ Guanella
Chris Hall
Darren Hall
Jeremy Hamm
Josh Hanfling
Savannah Hatcher
Hank Held
Gunnar Heuberger
Bill Hellman
Tim Hellman
Gunnar Heuberger
Kim Jackson
Joe Jankowski
Jordan Jankowski
Dan Johnson
Blayne Johnson
Dick Johnson
Ryan Johnson
William Johnson
Mike Jorgensen
Ralph Kalberloh
Barbara Kalberloh
Steve Keetch
Joe Keesee
Rex King
Jared King
Yale King
Bryan Knight
Don Knox
William Knowles
Melissa Kuipers
Kevin Lamar
Paula Lane

Jon Lind
David Liebowitz
Chris Lenckosz
Steve Lindner
Dean Littleton
Khorrie Luther
Len Lyall
Bill Maffeo
David Markley
Doug Markley
Gene Markley
Mike Maroone
Mike Marzolf
Phil Marzolf
Mike Marquez
Steve Maneotis
Tony Maneotis
Todd Maul
George McCadden
Mark McCadden
Scott McCandless
Alex McCandless
Corey McCandless
Zach McCandless
Robert McMann
Mark McMullin
George McCadden
Tammi McCoy
David McDavid
Doug McDonald
Jim McDonald
Michael McDonald
John Medved
Debbie Medved
Barron Meede
Gail Miller
Larry H Miller
Mark Miller
Jia Meeks
Aaron Mills
Jim Morehart
Brandon Moreland

Doug Moreland
Bonnie Murray
Pat Murphy
Steve Nilsson
Trent Olinger
Tom Ondrako
Brian O'Meara
Evan O'Meara
Paige O'Meara
Mike Nixon
Mary Pacifico
Nick Pacifico
Lee Payne
Jaymie Payne
Olga Payne
Michael Payne
Bob Penkhus
Mark Perez
David Perkins
Tom Perkins
Owen Perkins
Mark Peterson
Polly Penna
Gavin Pierce
George Pierce
Mitch Pierce
Steve Powers
Vic Reem
Ray Reilly
Tony Reinhart
Lannie Ridder
Greg Rowland
Justin Sasso
John Schenden
Kathy Schenden
Lisa Schomp
Vince Schreivogel
Jana Schreivogel
Mike Shaw
Michael Shaw
Susanne Shaw
Tom Sellers

RD Sewall
Kevin Shaughnessy
Scott Shimer
Jeff Silverberg
Amy Smith
Chris Smith
Bill Stahelin
Art Stapp
Brion Stapp
Bob Stapp
Clinton Stapp
Gregg Stone
Kent Stevinson
Paul Stevinson
Carol Spradley
Kriss Spradley
Larry Spradley
Jim Suss
Jim Suss, Jr.
Paul Suss
Wes Tabor

Steve Taylor
Jack TerHar
Paul Tew
Gary Tomkins
Linda Toteve
Joel Towbin
Ross Turner
Ed Tynan
Sean Tynan
Matt Tynan
Pete Tynan
Mike Tynan
Tim Van Binsbergen
Carl Venstram
Barbara Vidmar
Bill Vidmar
Derek Vidmar
Aaron Wallace
Mark Wallace
Bill Walters
Mike Ward

Patrick Watson
Jay Weibel
Roger Weibel
Peter Welch
Tim Wieland
Bill Wilcoxson
Ken Williams
Tyrone Williams
Dan Wilson
Phil Winslow
Ann Winslow
Dan Wilson
Bill Wolters
Brent Wood
Judi Wood
Chase Yoder
Lee Yoder
Warren Yoder
Steve Zeder
Mark Zeigler

(I completed the above list entirely by memory—sorry for any important omissions).

Sketch artist credits:
- Engel Morales
- Nishani Sanjeewani
- Mike Parlenko
- Vashishtha Ghodasara
- Fakker Ul Islam

Cover time lapse photo credit
- Michael Ryno

Book Designer:
- Nick Zelinger, NZ Graphics

Audiobook:
- Richard Rieman, Image History Books

About the Author

The multitalented Tim Jackson doesn't sit around and wait for others to set the agenda.

He's been a licensed private pilot since age 19. He considers himself a car guy, a plane guy and a bike guy—cycling between 2,000 and 4,000 miles a year and completing 16 cross-the-state bike tours of Ride the Rockies, Colorado's premier cycling event and a similar ride across Iowa called RAGBRAI.

Tim's 35-year career in association management demonstrated great leadership abilities promoting activism in government and politics. From organizing a nationwide 48-state, 1600 event, bus tour to activate young people in saving Social Security for future generations to creating a non-profit foundation that targeted high emitting sources of air pollution and using the value of donated contributions for scholarships for those entering mobility industries, Tim has a reputation and track record as a proven solid goal achiever.

Tim is a consistent promoter of the automobile industry and cherishes and values the freedom it provides. As the principal producer of the Denver Auto Show for 18 years, Tim was able to expand it to double revenue, add a preview gala, include a Colorado Automotive Hall of Fame and create an Innovative Dealer Summit.

Tim collaborated with Anthony Brownlee to create the hall of fame that has now inducted over 65 top industry leaders who excelled exceptionally in their automotive careers.

Tim believes in promoting the industry in the direction it needs to go. Including one company car, Tim has already had five electric cars—and currently has two—while others are still trying to figure out how to electrify their garages. Tim is a strong believer that we'll all soon be flying cars. Don't believe him? Ask to see his spectacular presentations. He'll be happy to show you.

Tim elevated the profile and stature of his assigned associations in the public eye. He increased activation and participation of membership throughout the state, with increased emphasis on member meetings and events. Several of the associations Tim led grew their net assets by more than double.

The organizations Tim has led have increased in value, attracted new members, elevated their community and public profile, and grown in effectiveness. In successive organizations, Tim led grassroots organizing programs that provided impressive member engagement in the political process and ultimate success in the legislative and advocacy results.

Tim believes in the freedom, power and importance of personal choice in transportation and mobility. Through direct affiliations, Tim argued for vehicle choice and personal choice of power source technology.

Tim has seen a lot of changes and advancements in the automotive world, but he is really excited about improvements in overall vehicle quality, varied power source technology and advanced autonomous drive technology. He is an ambassador for such concepts. He spreads his enthusiasm about the car business, whether it's new designs, new concepts in the field or the news from one of the many companies that hope to create and take over a future market of flying cars.

Tim is widely known for being tenacious, energized, informed and activated. His typical workday starts before 6 a.m. and ends at or after 10 p.m. Tim never met a stranger. He can work a room full of friends better than most Members of Congress. By the end of the gathering, Tim will have met everyone in the room. This personal and up-front approach has benefited Tim in his public policy advocacy efforts.

Crack open this book and learn about the mobility that drives this man of so many interests. Learn about what's coming at us fast, whether it's flying cars, Tim on a challenging bicycle ride in the middle of the Colorado mountains, or how to promote programs and convince people to join associations that match their interests.

See Tim on the road—at least until cars start flying. Then wave at him when passing in the sky.

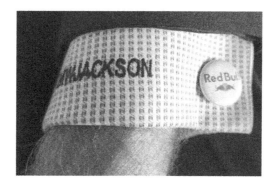

Dear Reader,
I hope you enjoyed reading
Dude, Where's My Flying Car?
I would greatly appreciate you
posting a review on my Amazon page.

Book Tim Jackson for your next engagement

timwjacksoncae@gmail.com

303-667-3995

Made in the USA
Monee, IL
16 August 2024

63382554R00272